KS2 Success

Age 7-11

Maths

Trevor Dixon

Tests

Contents

Arithmetic

TEST 1

- You have 20 minutes to complete this test.
- Calculator not allowed.
- Use the spaces provided for your workings. Where two marks are available, you may be awarded a mark for your workings.

1 479 + 100 =

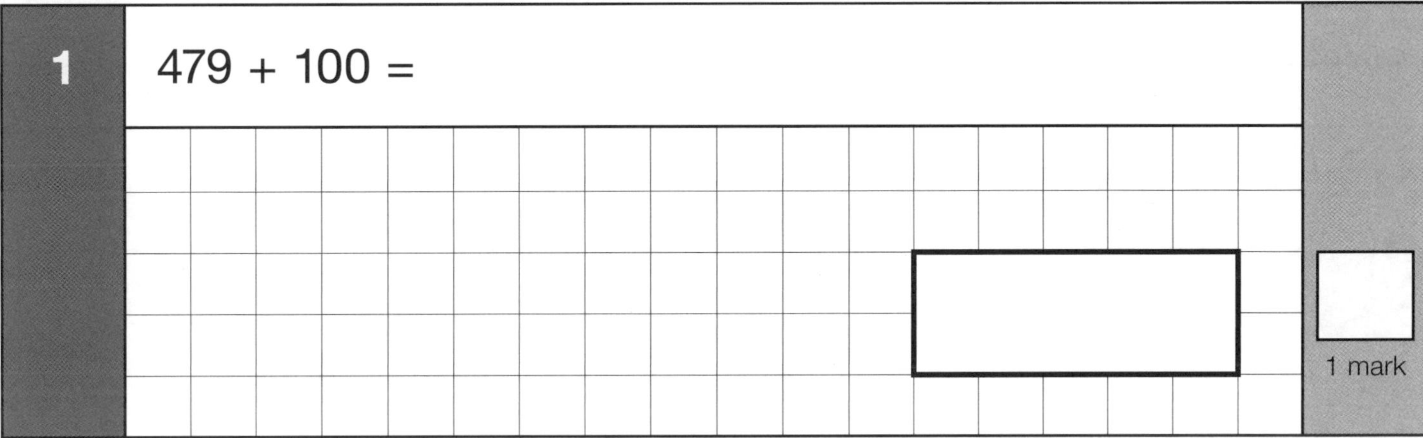

2 7256 + 847 =

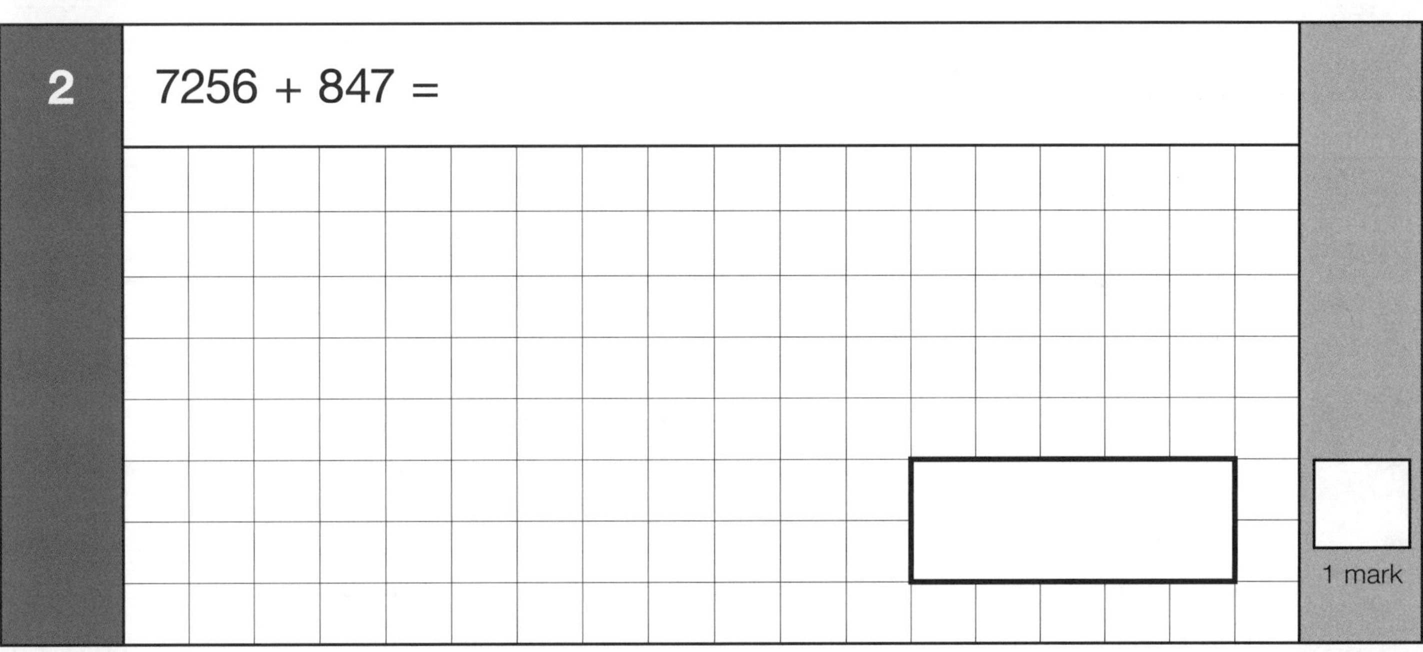

3 $\frac{9}{10} + \frac{7}{10} =$

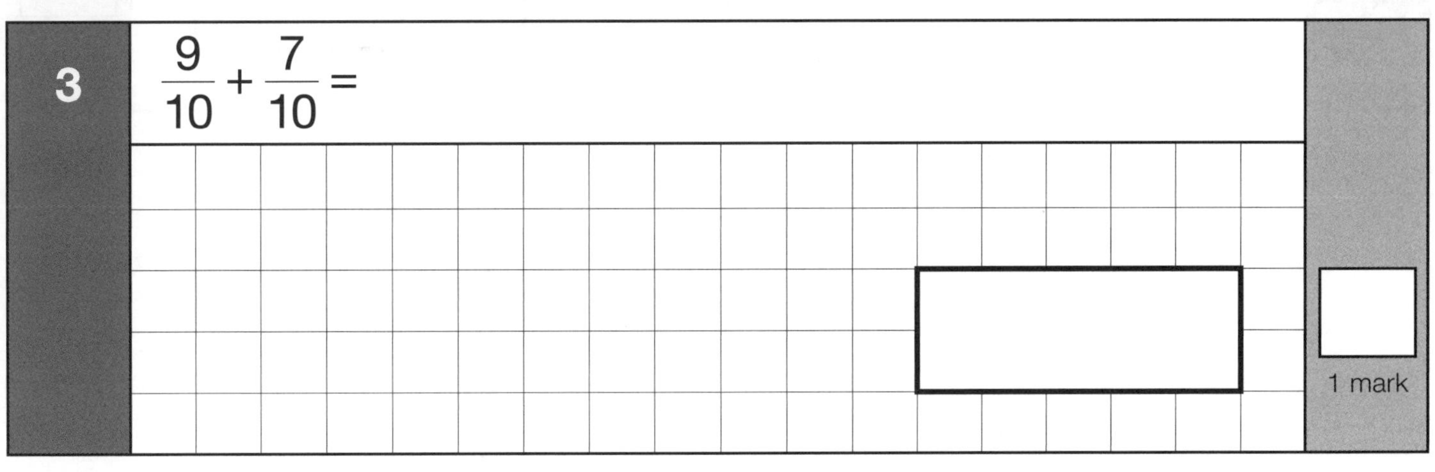

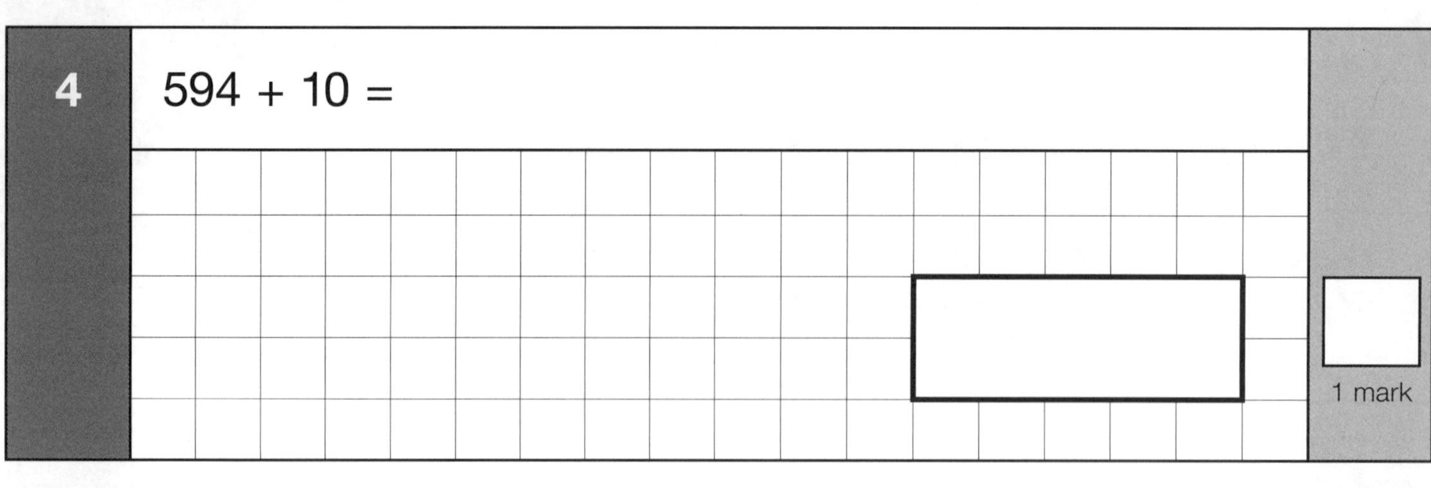
4
594 + 10 =
1 mark

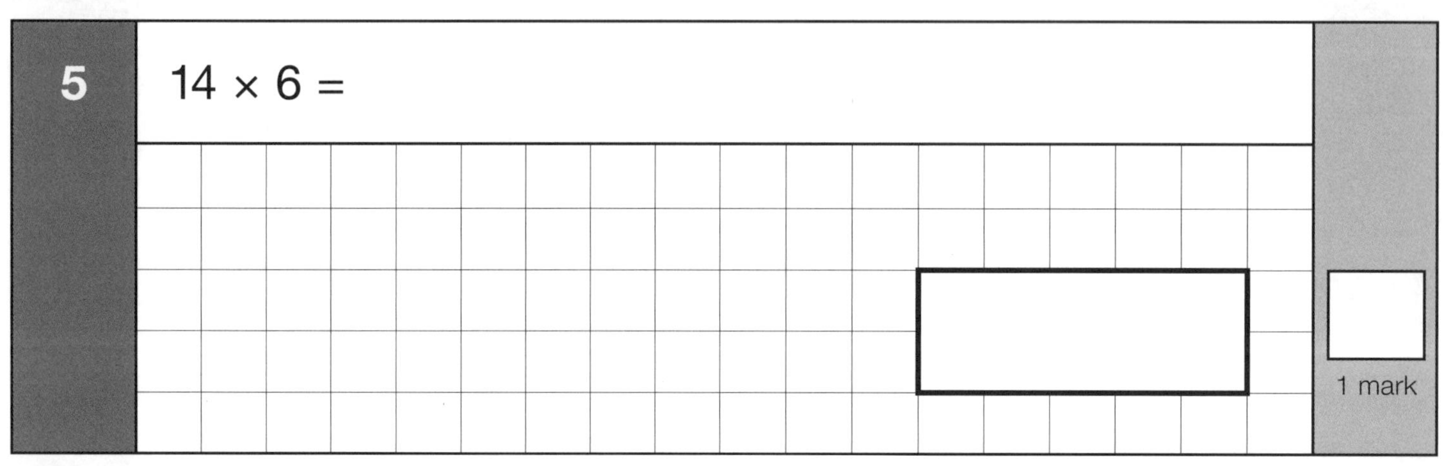
5
14 × 6 =
1 mark

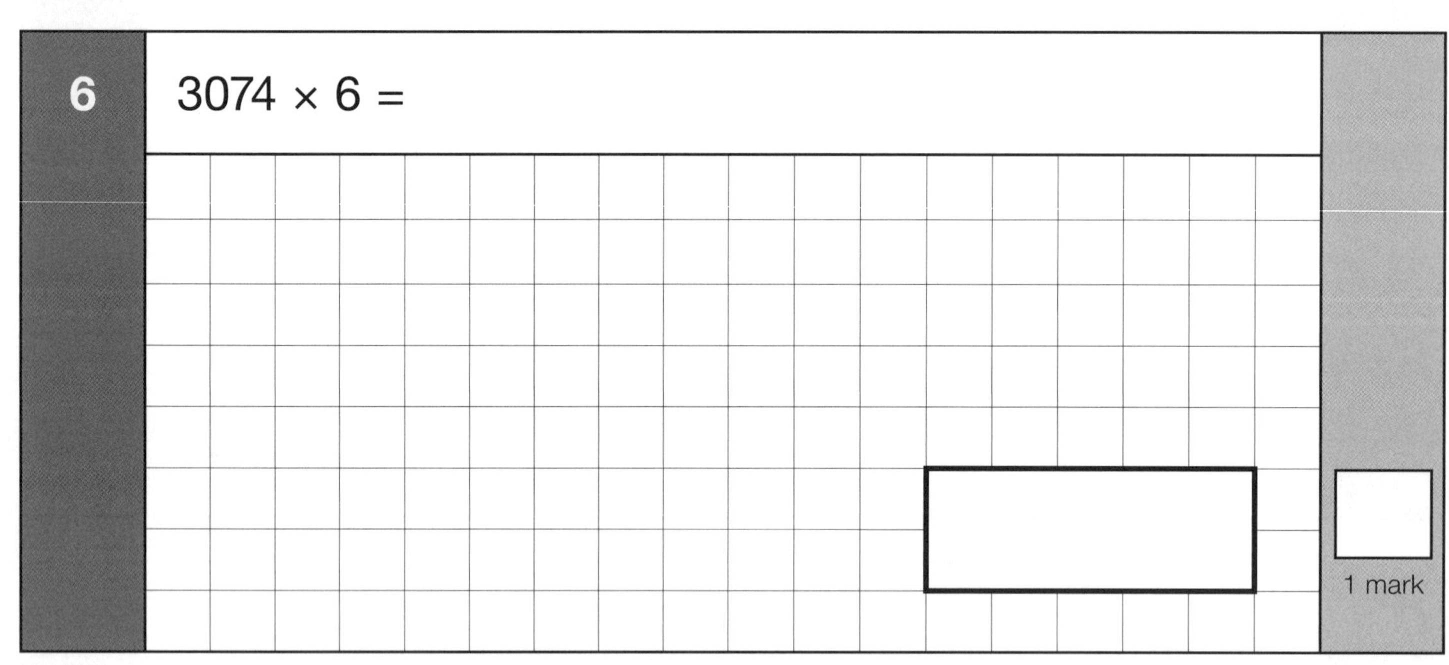
6
3074 × 6 =
1 mark

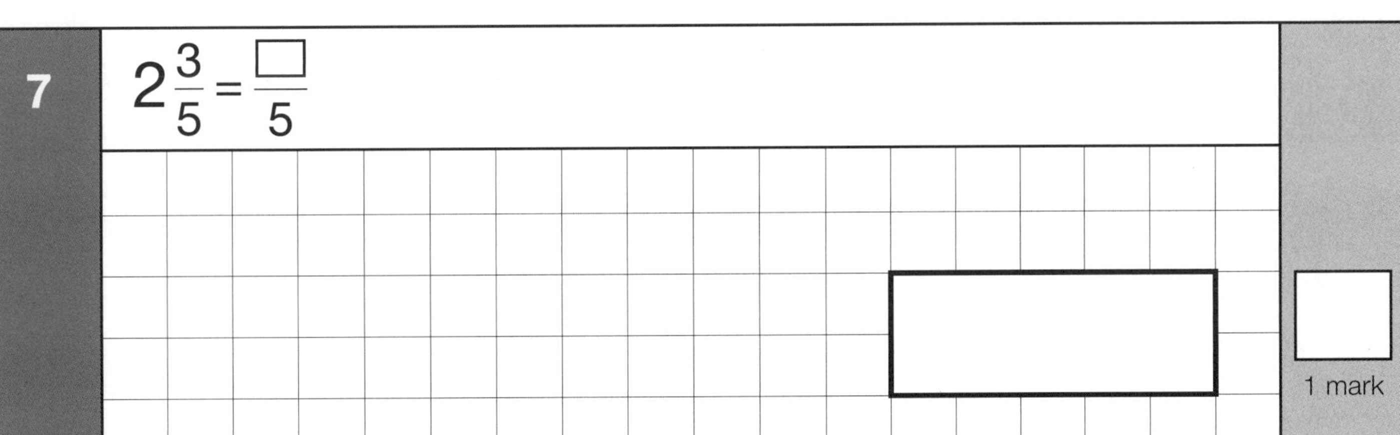

7 $2\frac{3}{5} = \frac{\square}{5}$

1 mark

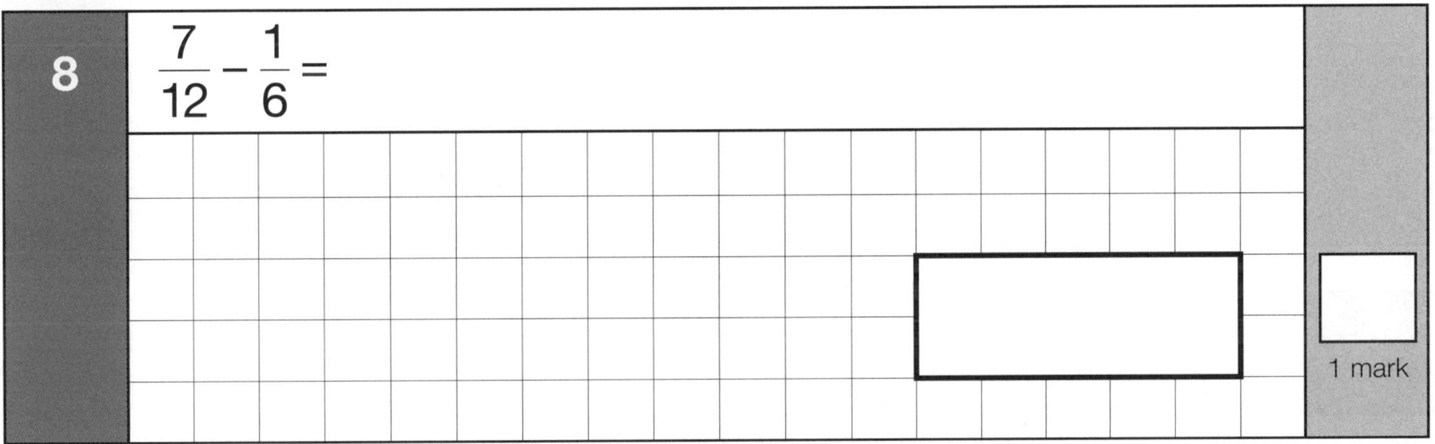

8 $\frac{7}{12} - \frac{1}{6} =$

1 mark

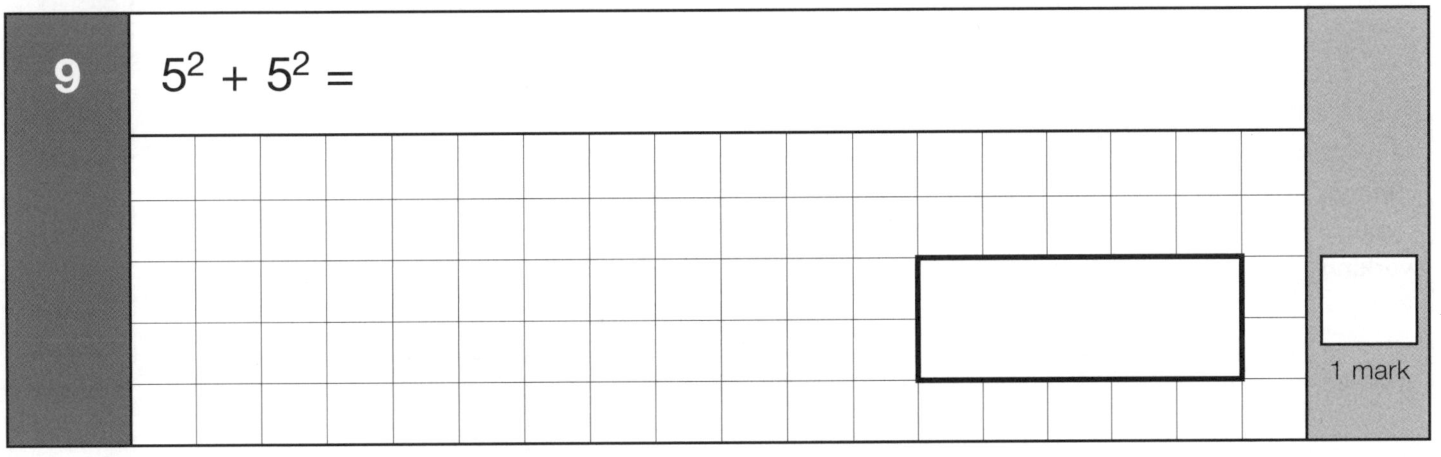

9 $5^2 + 5^2 =$

1 mark

10

$$\begin{array}{r} 328 \\ \times \quad 26 \\ \hline \end{array}$$

Show your working

2 marks

11

$$25 \overline{)1775}$$

Show your working

2 marks

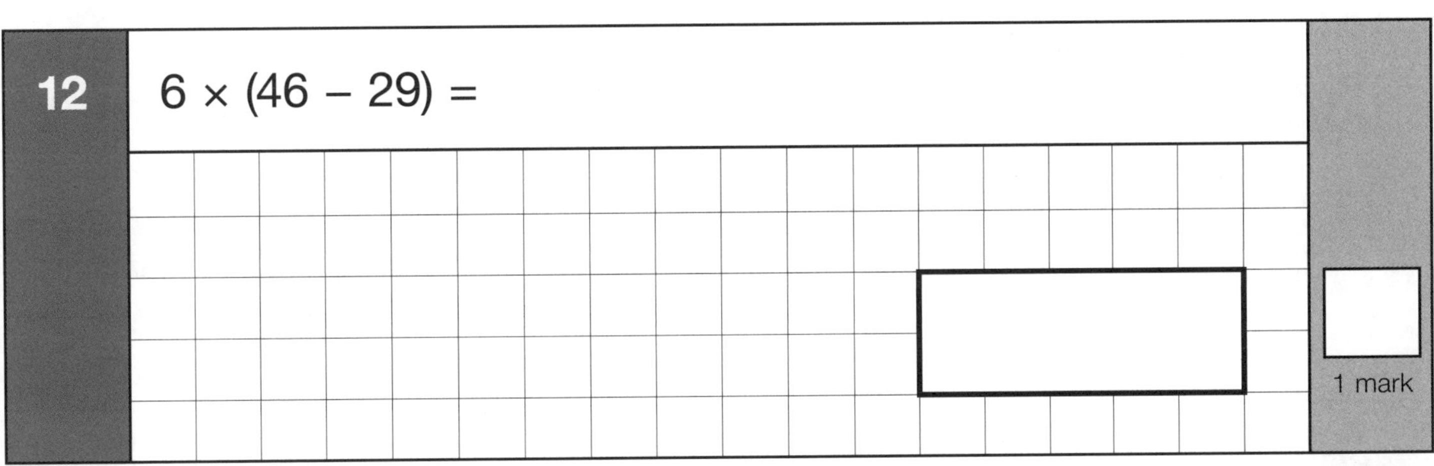

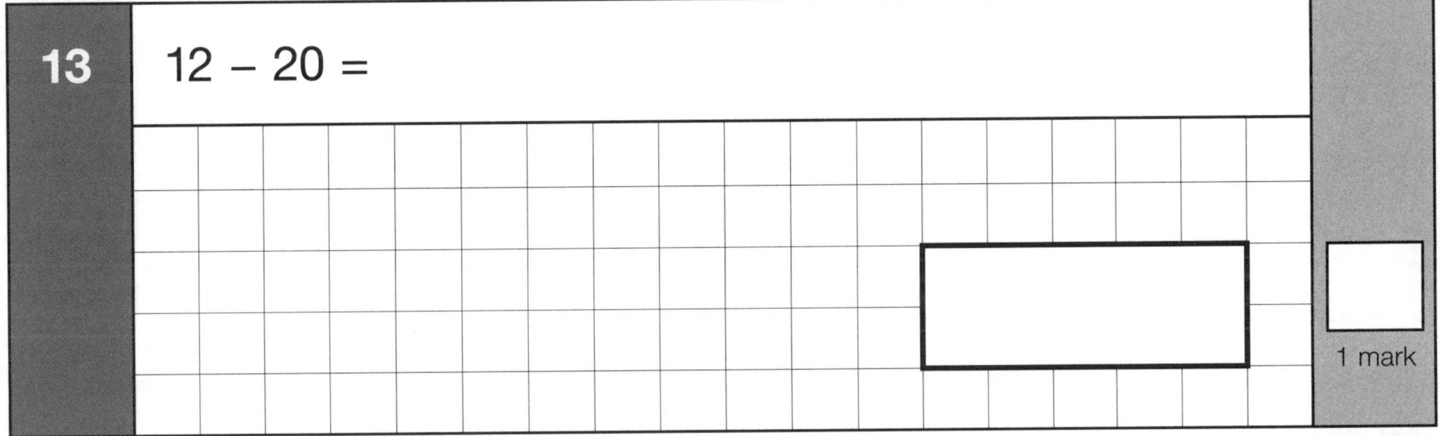

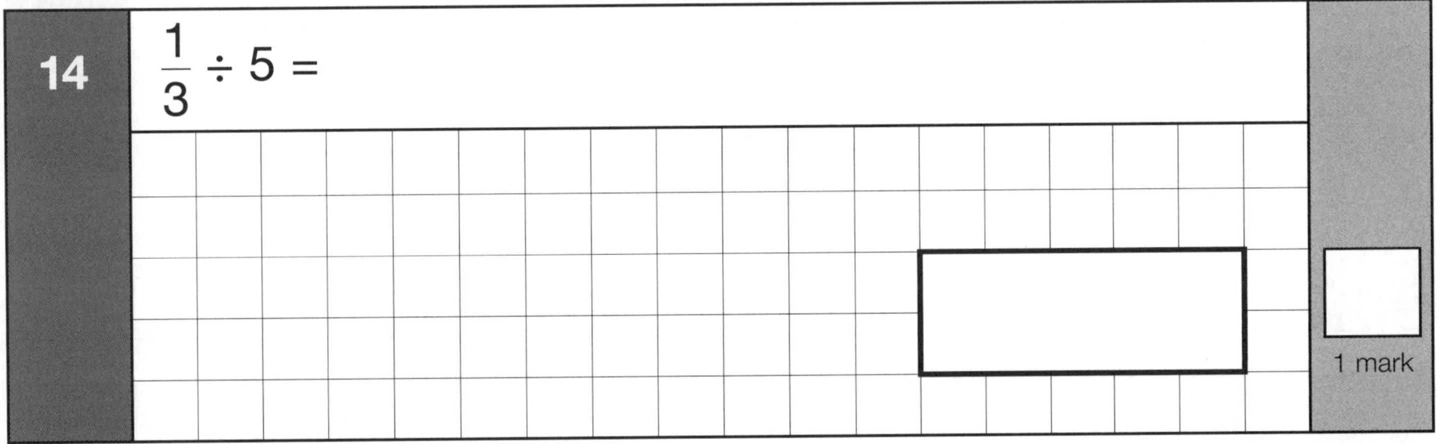

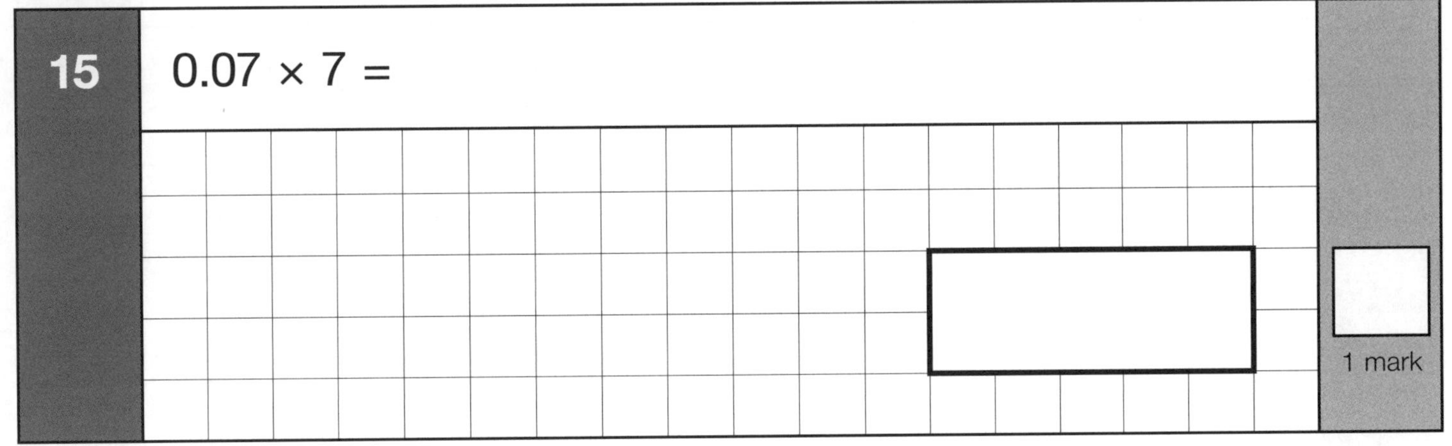

Total ________ / 17 marks

TEST 2

Arithmetic

- You have 20 minutes to complete this test.
- Calculator not allowed.
- Use the spaces provided for your workings. Where two marks are available, you may be awarded a mark for your workings.

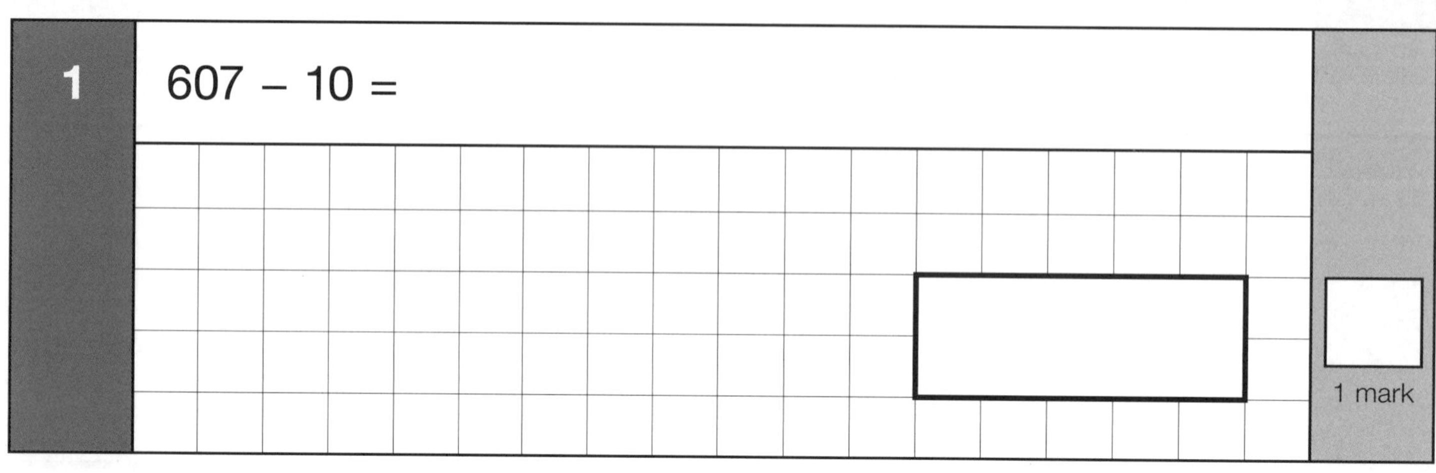

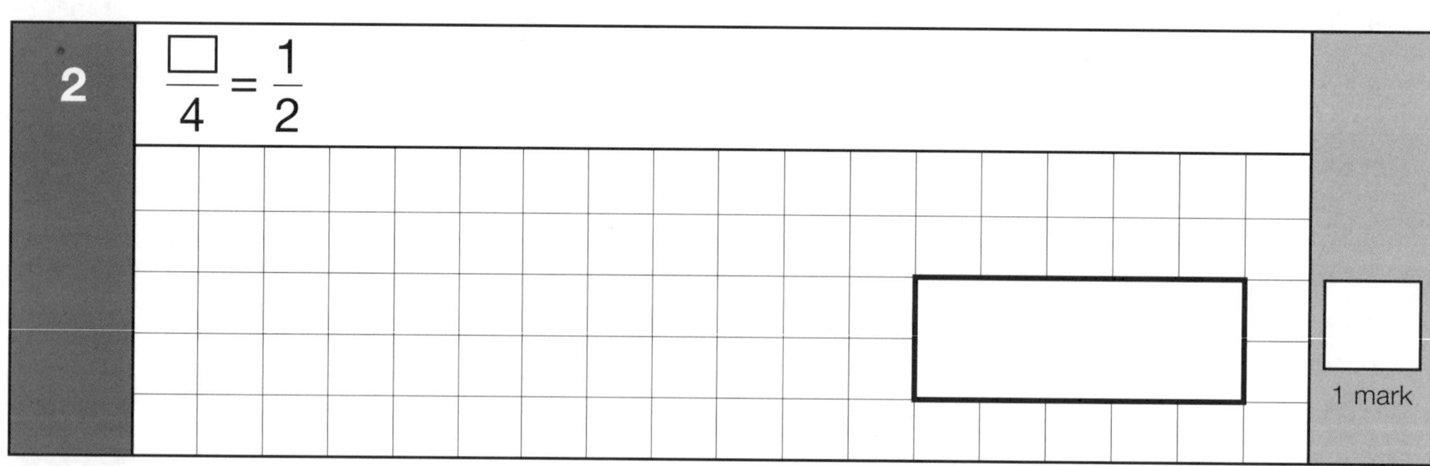

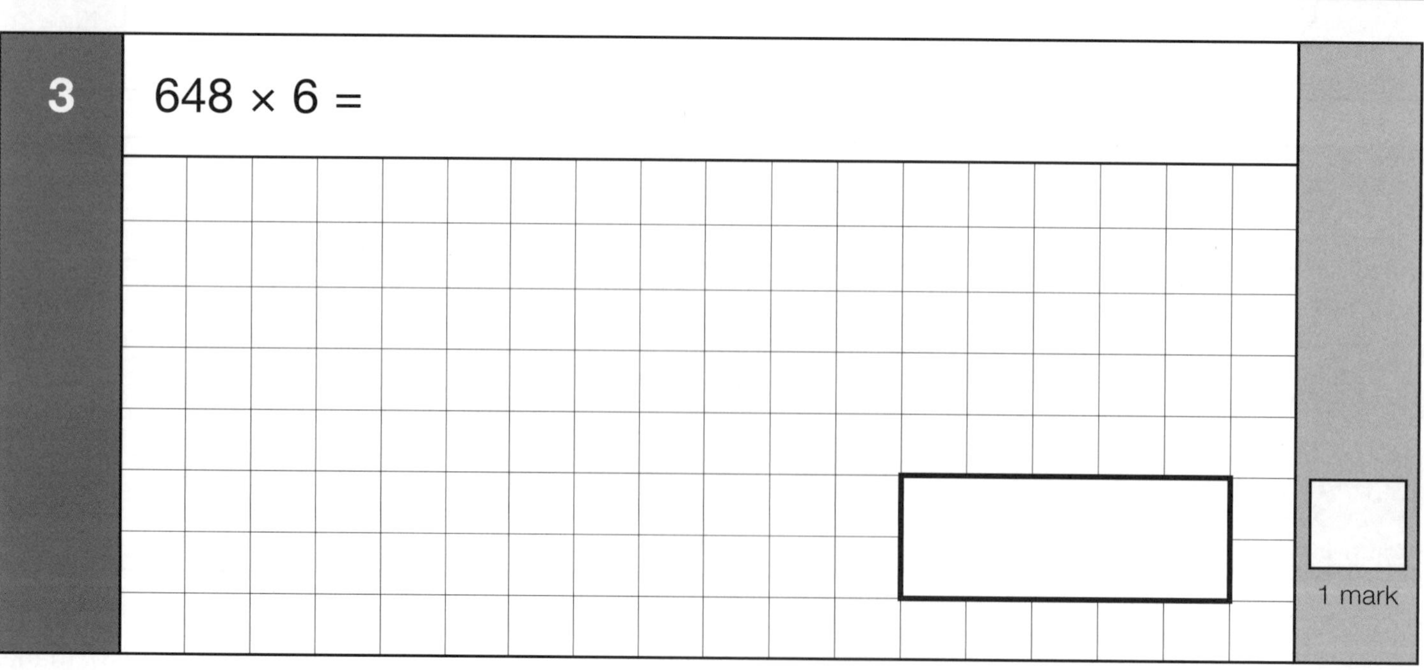

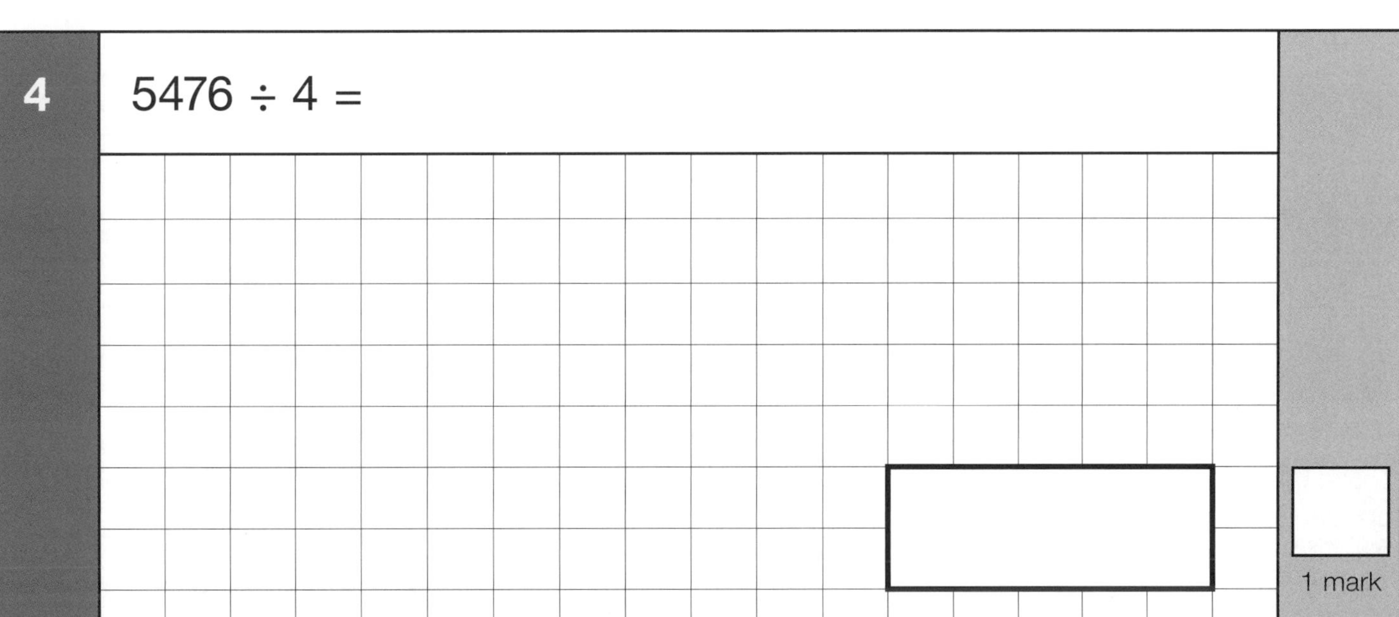
4
5476 ÷ 4 =
1 mark

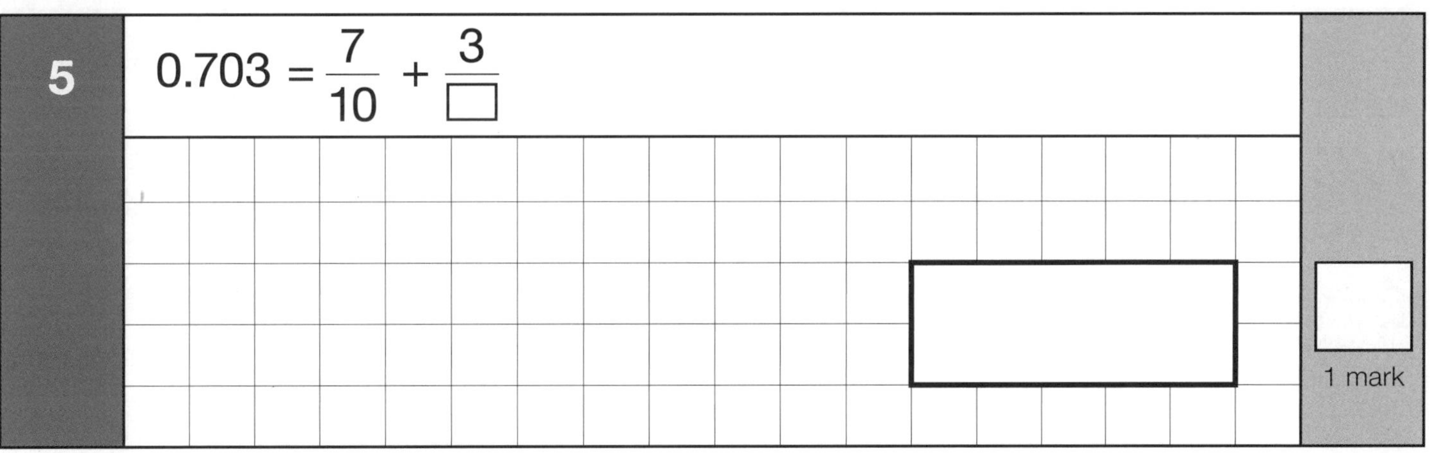
5
0.703 = 7/10 + 3/□
1 mark

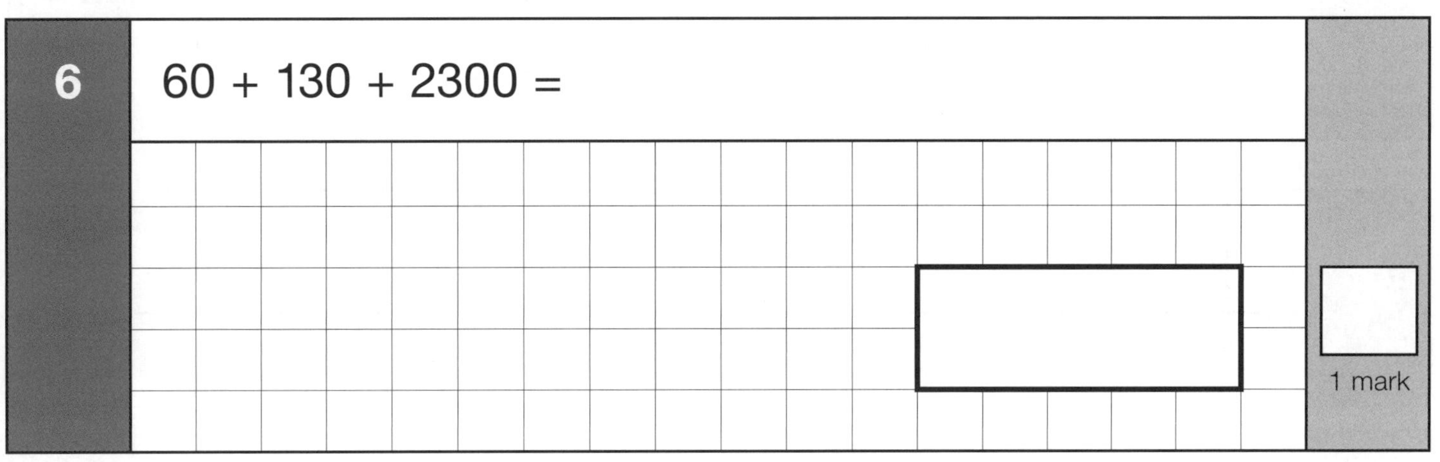
6
60 + 130 + 2300 =
1 mark

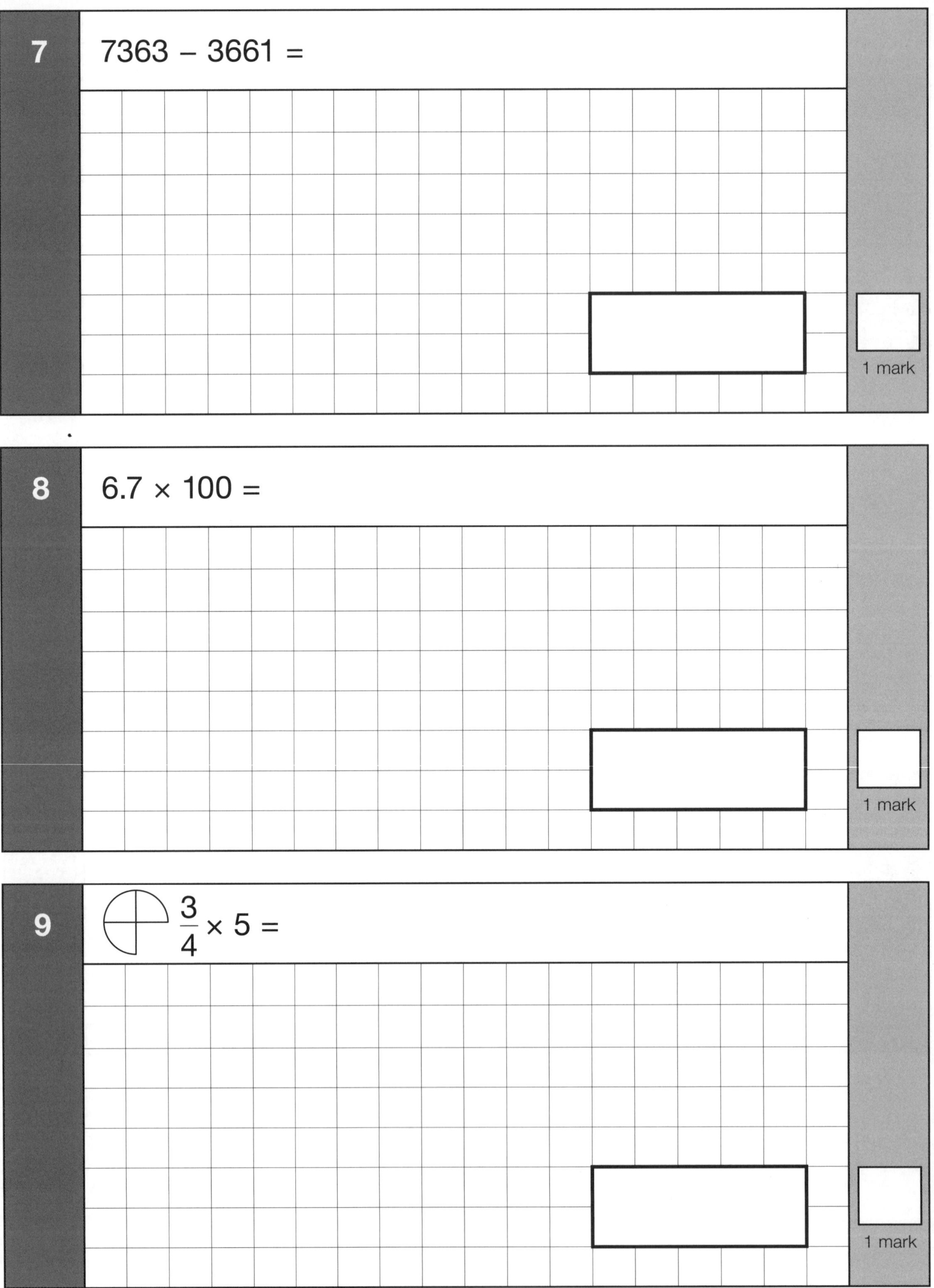
7
7363 − 3661 =
1 mark
8
6.7 × 100 =
1 mark
9
$\frac{3}{4} \times 5 =$
1 mark

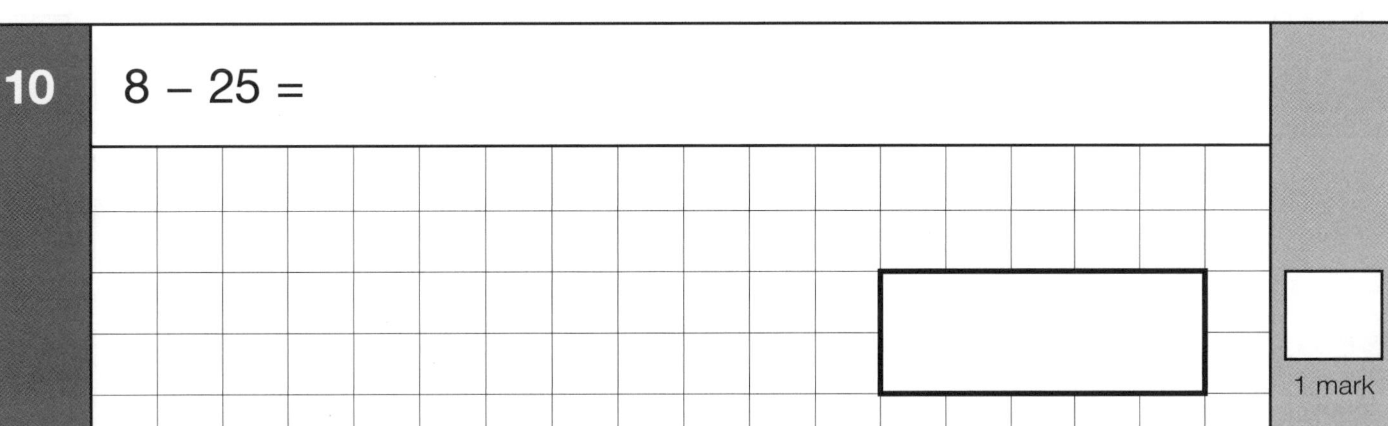
10
8 – 25 =
1 mark

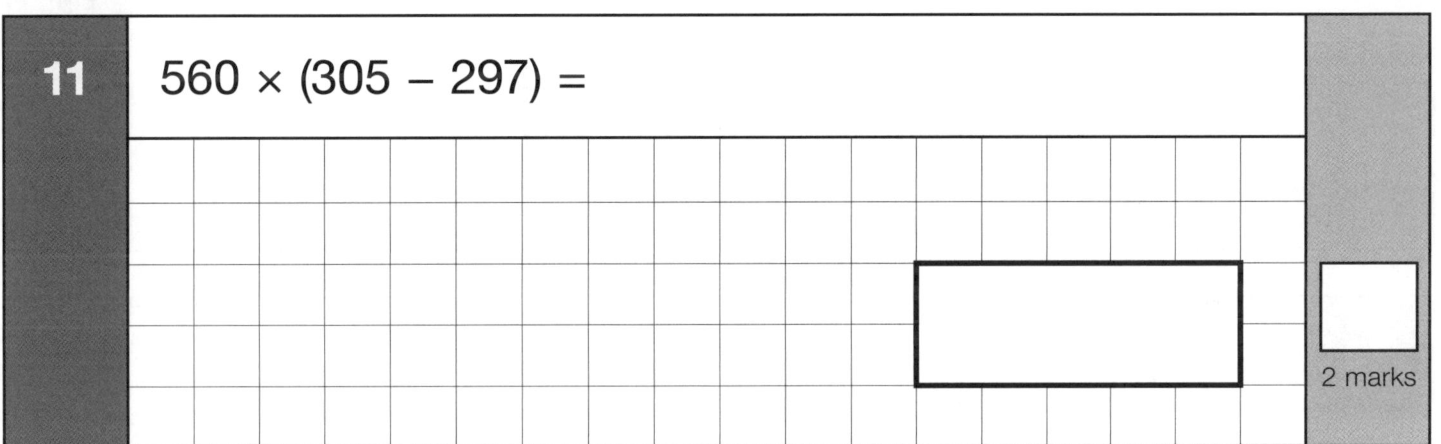
11
560 × (305 – 297) =
2 marks

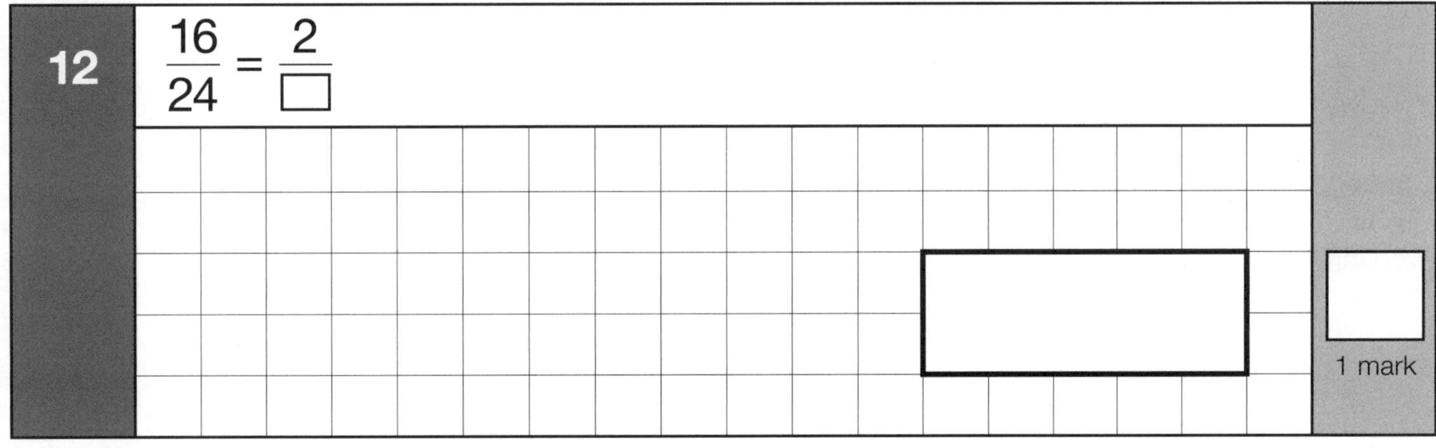
12
16/24 = 2/□
1 mark

13

$$\begin{array}{r} 6\,4\,8 \\ \times \quad 3\,5 \\ \hline \end{array}$$

Show your working

2 marks

14

$$32\overline{)5792}$$

Show your working

2 marks

Total ________ / 17 marks

Arithmetic

TEST 3

- You have 20 minutes to complete this test.
- Calculator not allowed.
- Use the spaces provided for your workings. Where two marks are available, you may be awarded a mark for your workings.

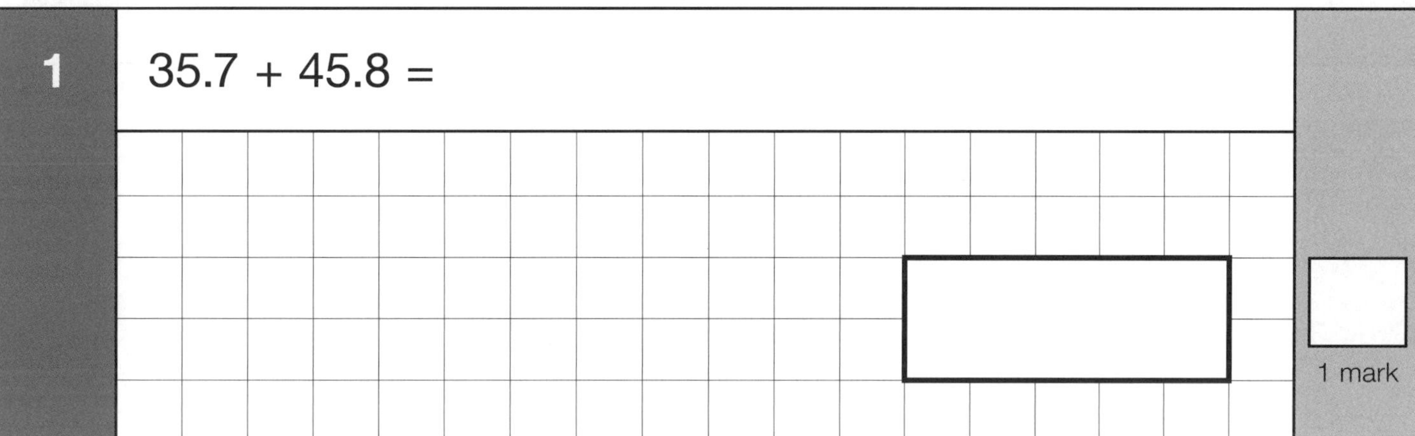

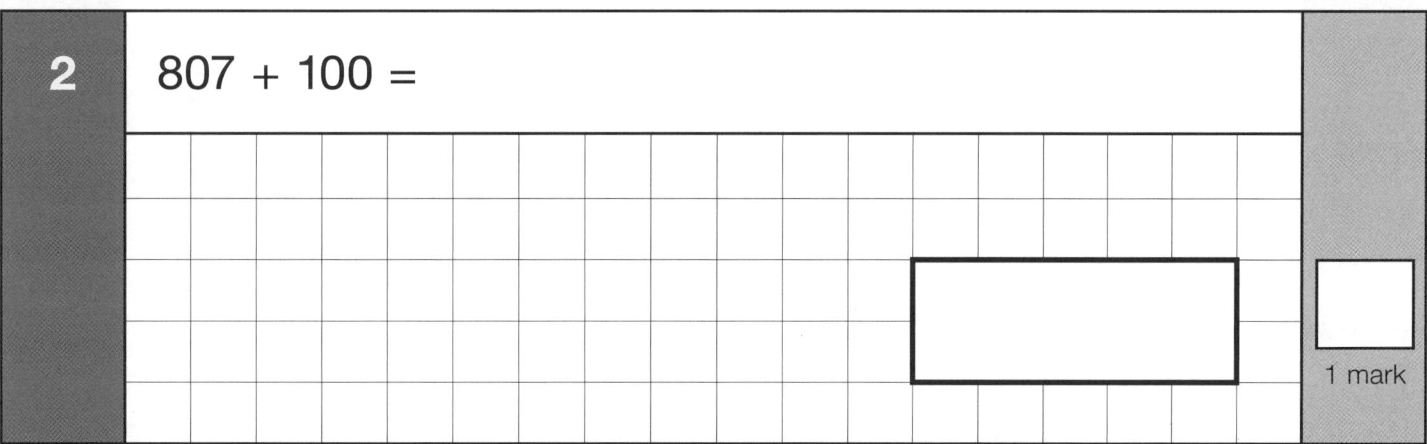

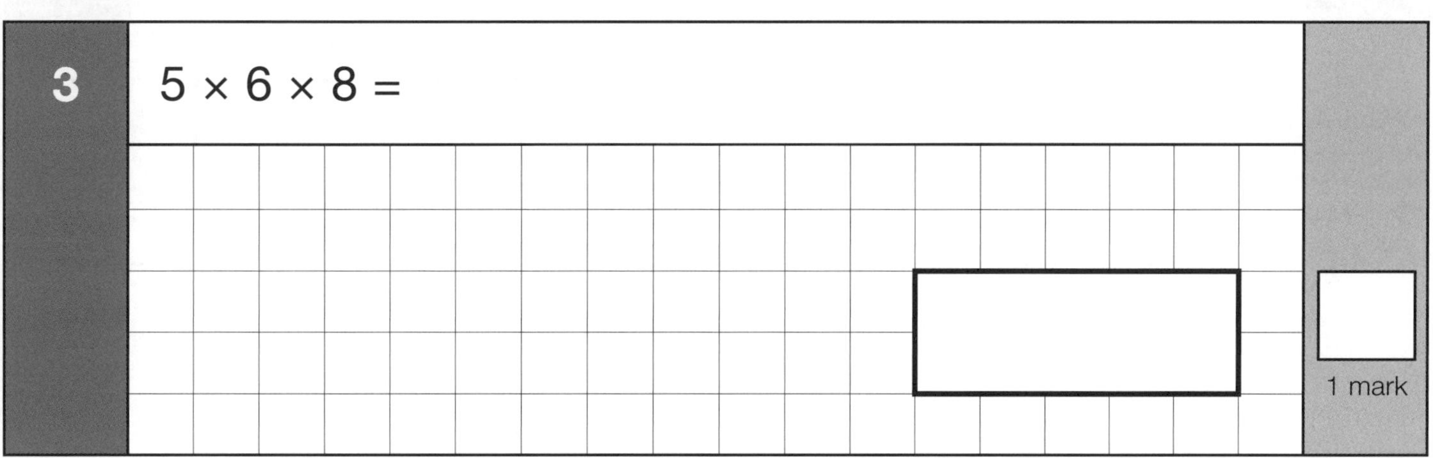

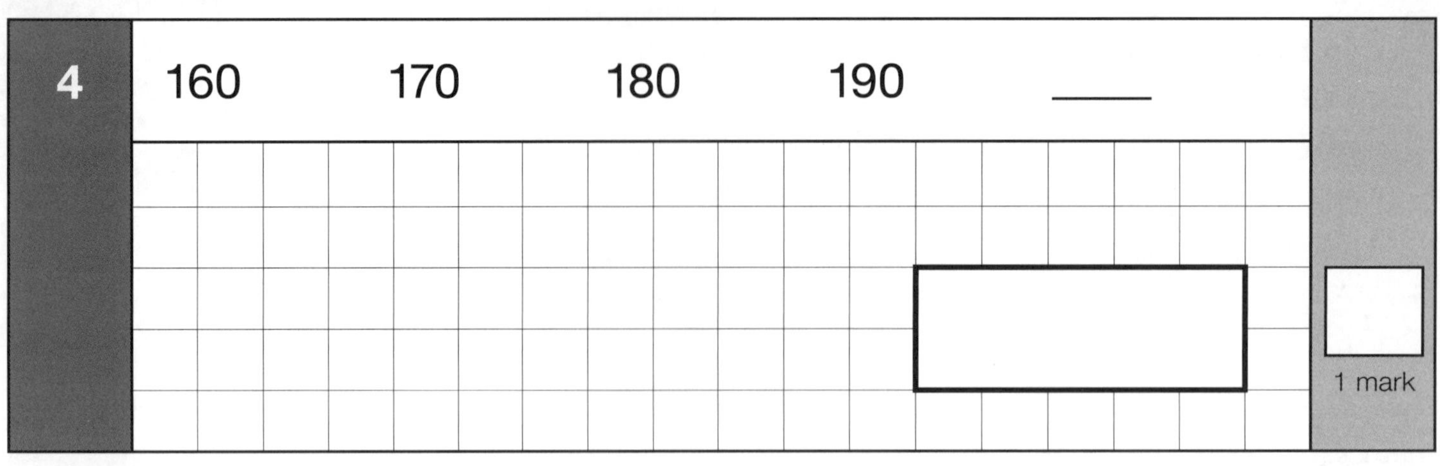
4
160 170 180 190 ____
1 mark

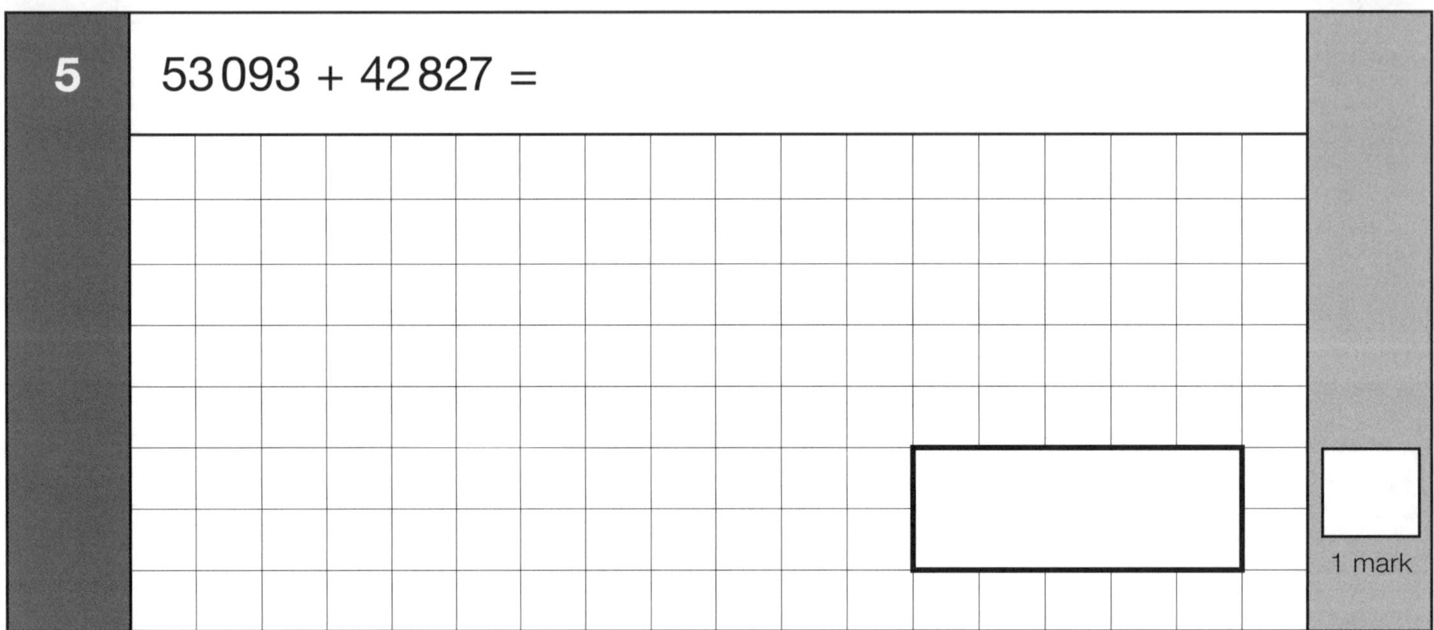
5
53 093 + 42 827 =
1 mark

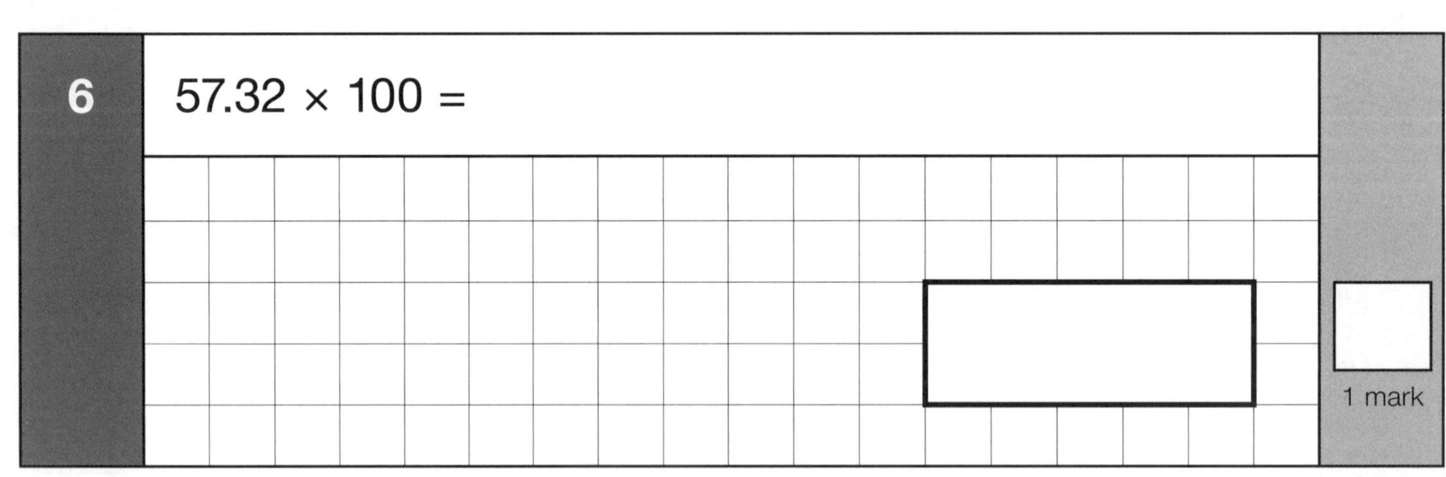
6
57.32 × 100 =
1 mark

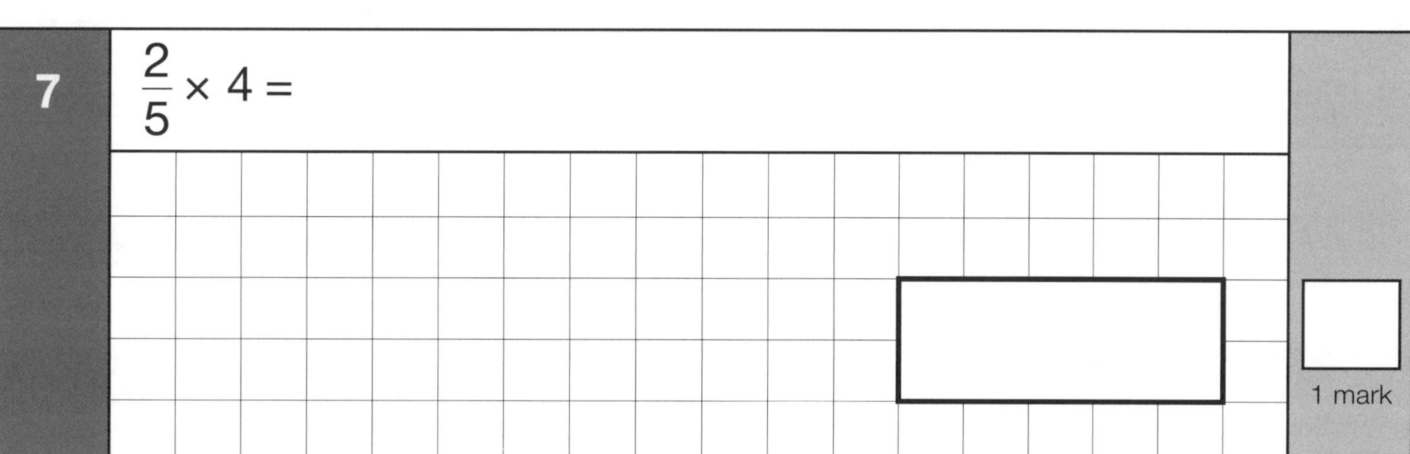

7 $\frac{2}{5} \times 4 =$

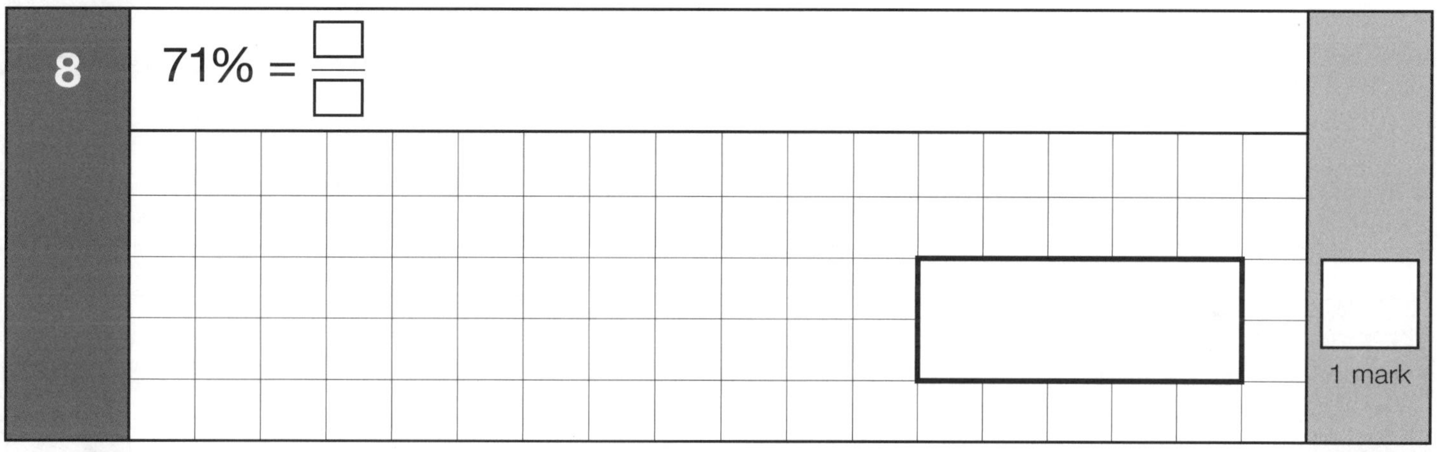

8 $71\% = \frac{\square}{\square}$

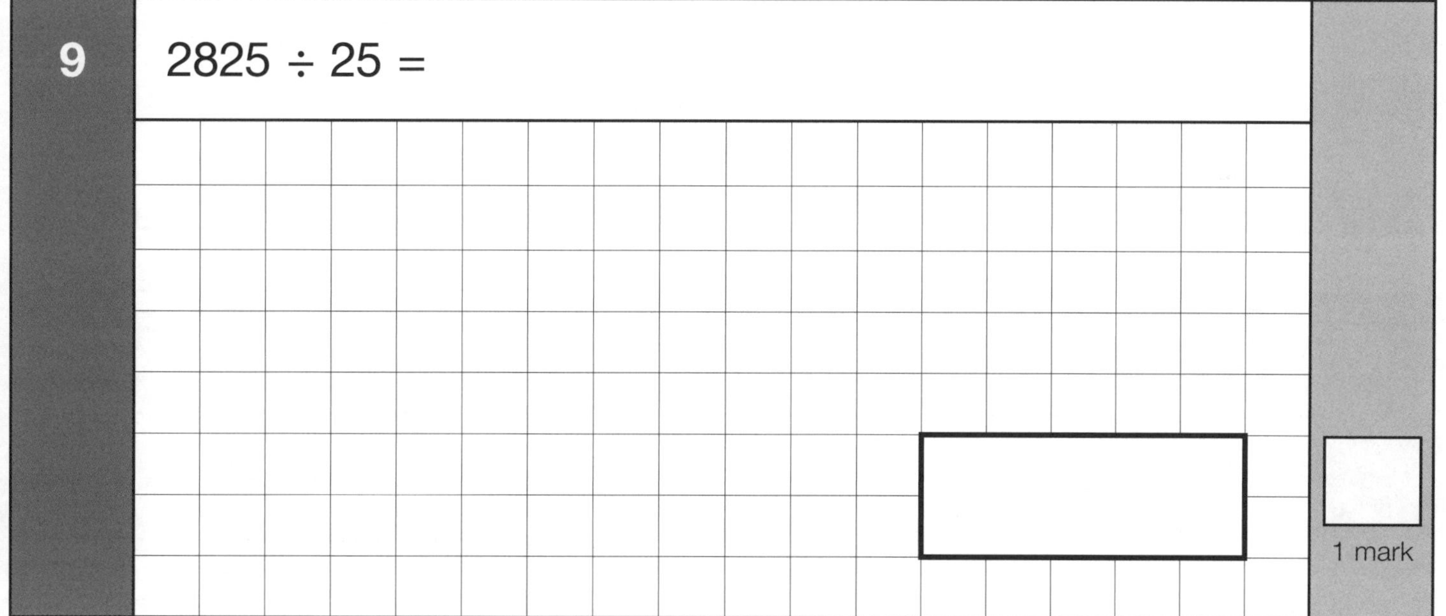

9 $2825 \div 25 =$

10

$$\begin{array}{r} 506 \\ \times \quad 52 \\ \hline \end{array}$$

Show your working

2 marks

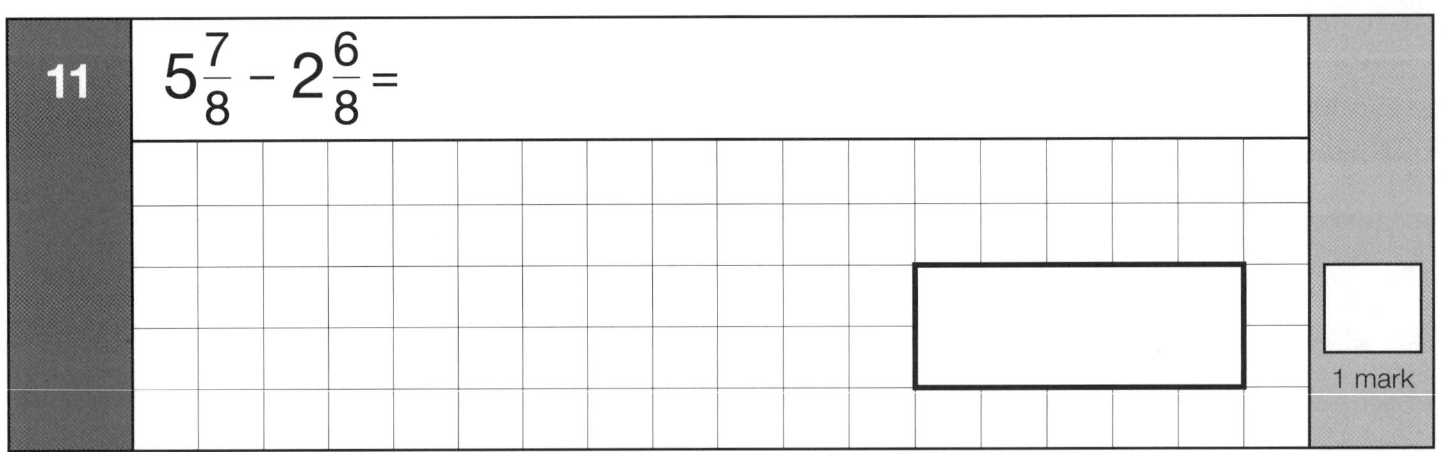

11 $5\frac{7}{8} - 2\frac{6}{8} =$

1 mark

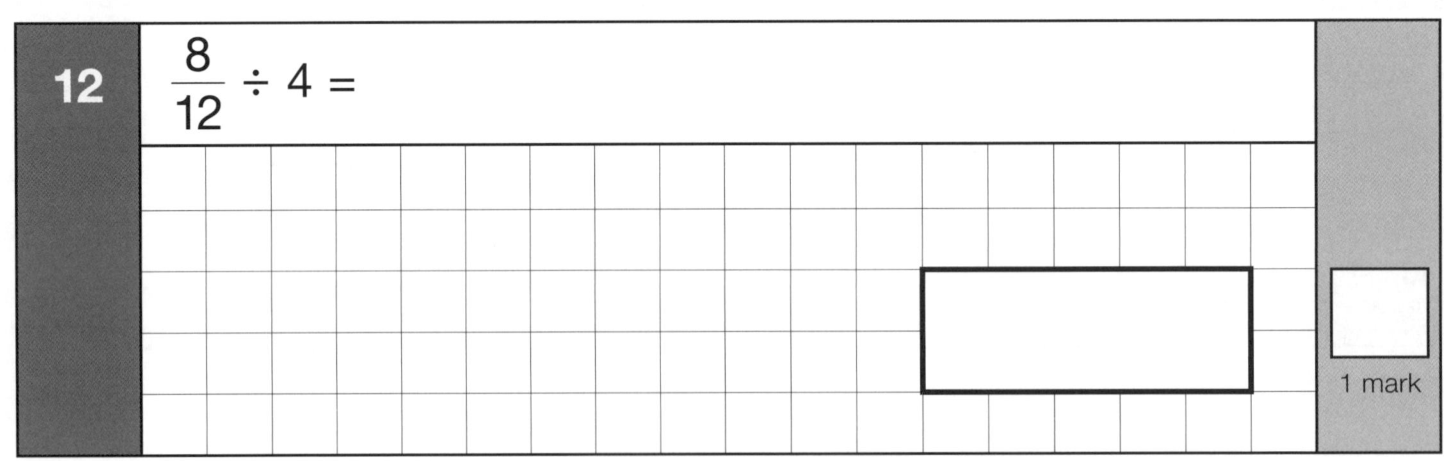

12 $\frac{8}{12} \div 4 =$

1 mark

13

$$32\overline{)768}$$

Show your working

2 marks

14

$(304 - 160) \div 24 + 24$

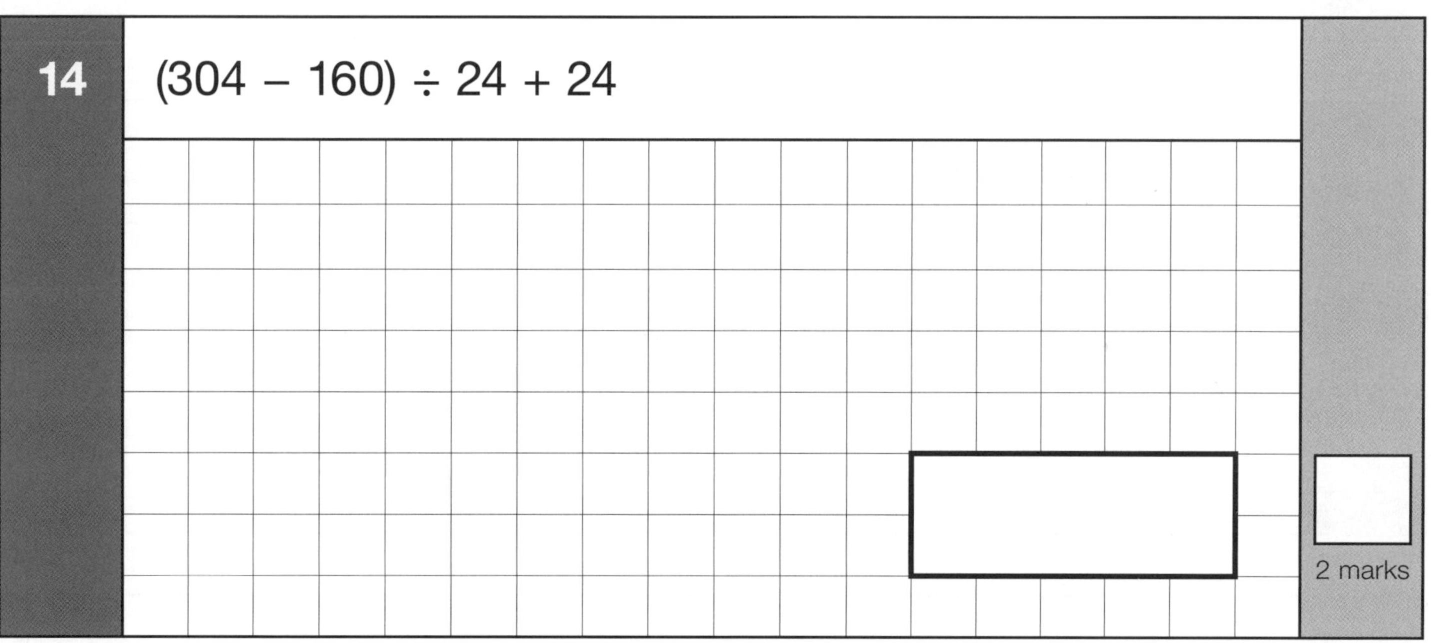

2 marks

Total ________ / 17 marks

TEST 4

Arithmetic

- You have 20 minutes to complete this test.
- Calculator not allowed.
- Use the spaces provided for your workings. Where two marks are available, you may be awarded a mark for your workings.

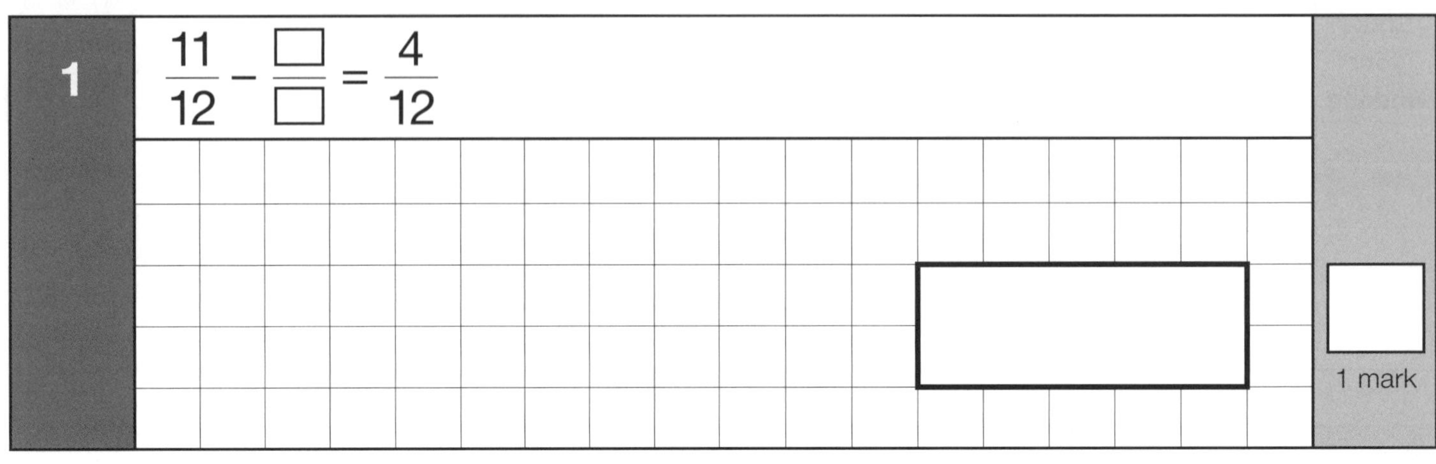

1 $\frac{11}{12} - \frac{\square}{\square} = \frac{4}{12}$

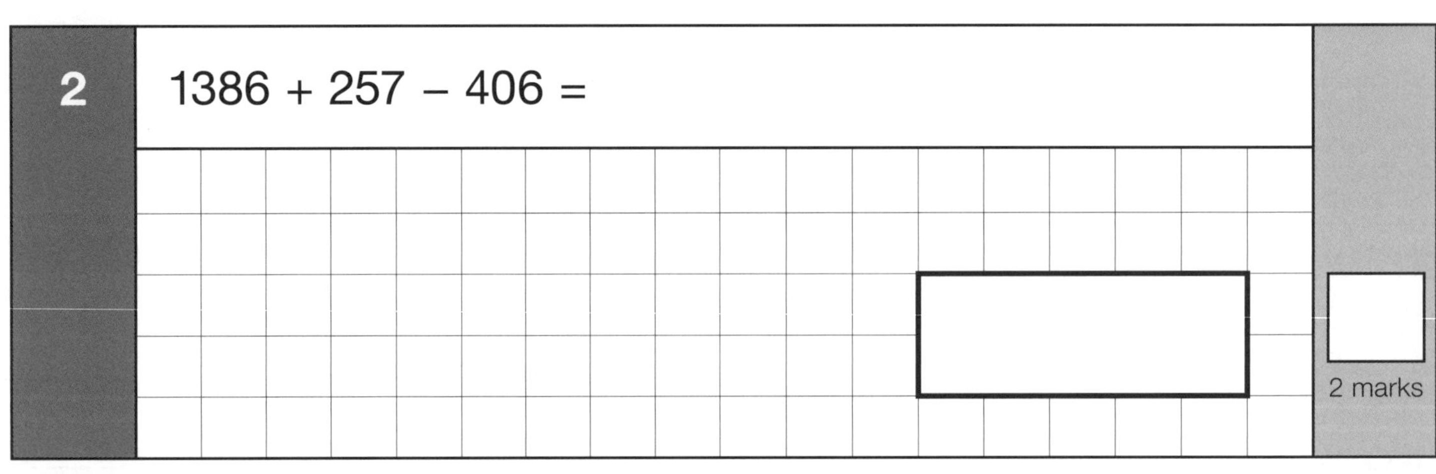

2 $1386 + 257 - 406 =$

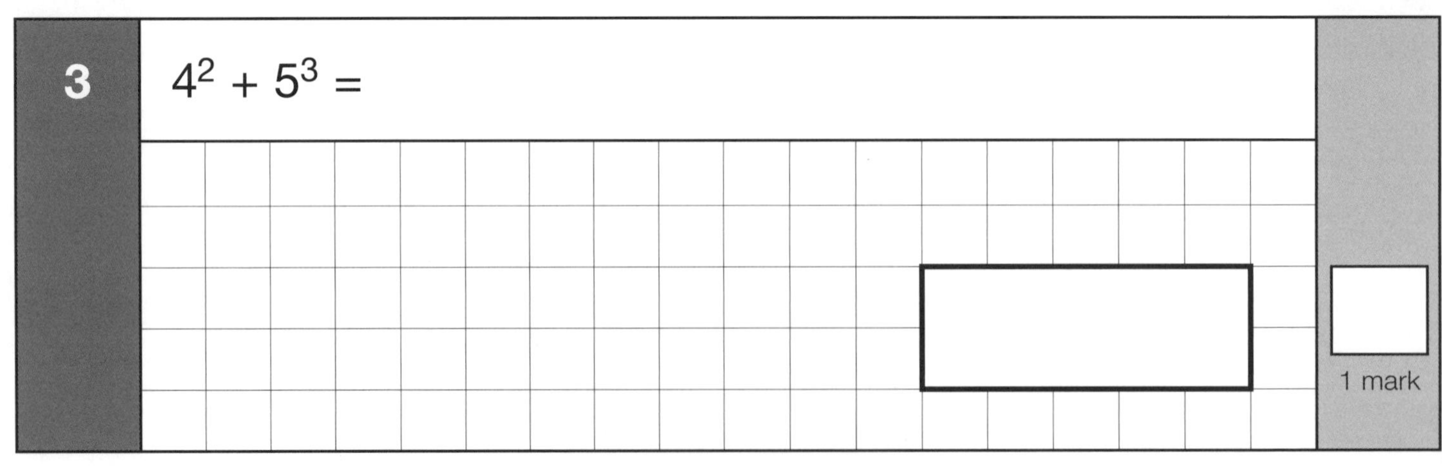

3 $4^2 + 5^3 =$

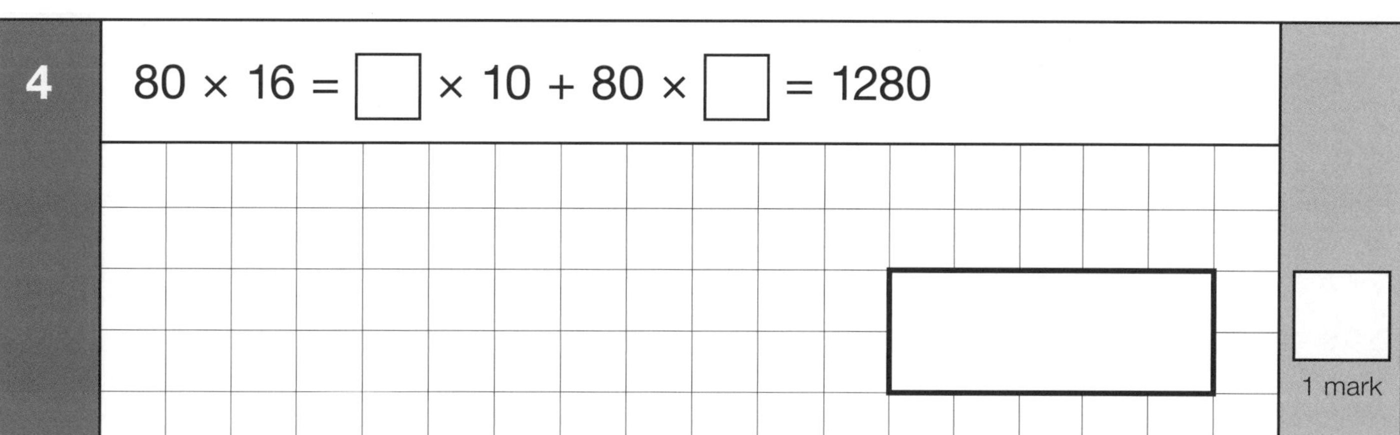
4
$80 \times 16 = \square \times 10 + 80 \times \square = 1280$
1 mark

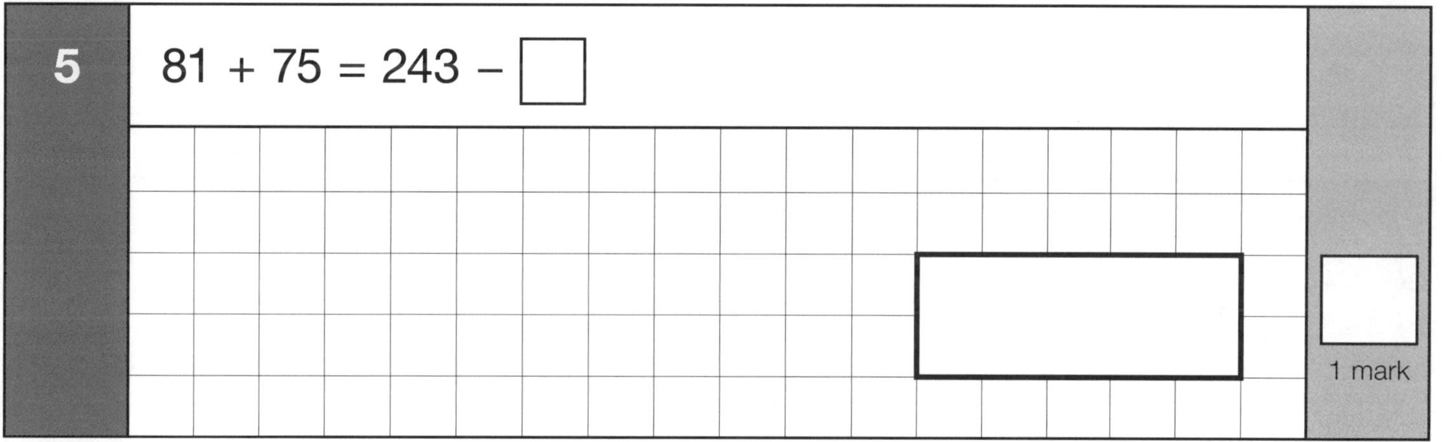
5
$81 + 75 = 243 - \square$
1 mark

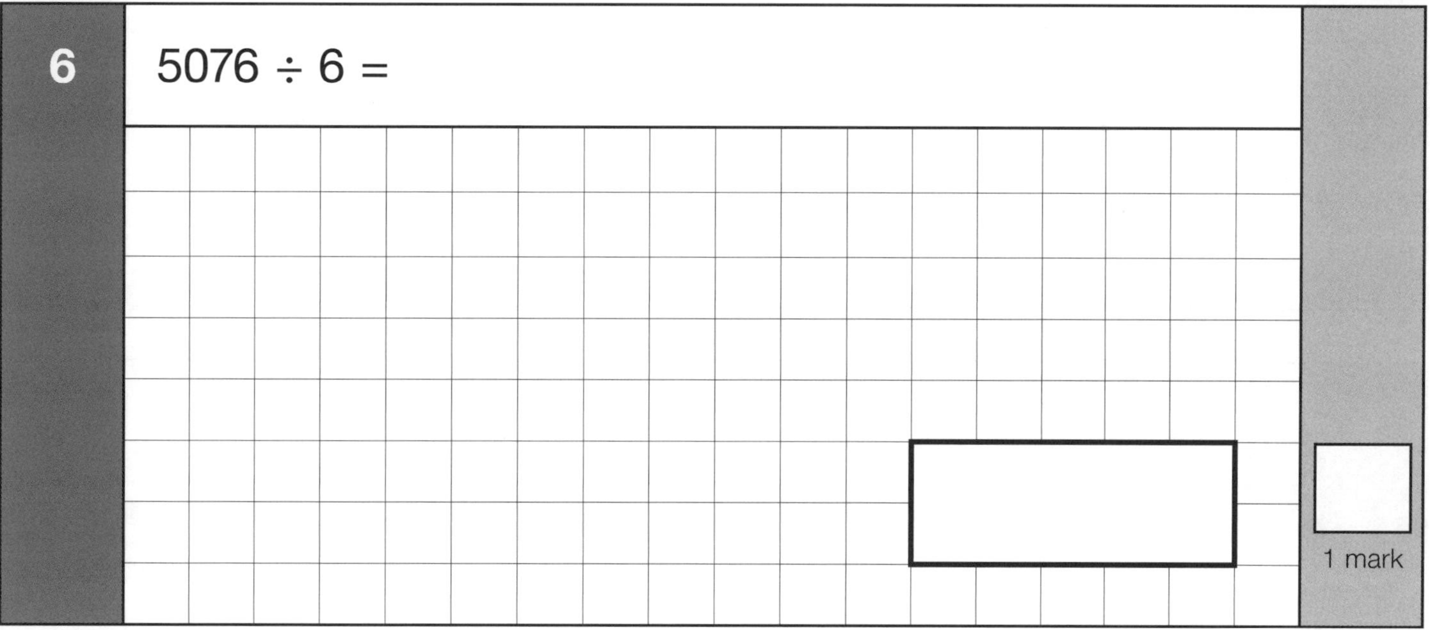
6
$5076 \div 6 =$
1 mark

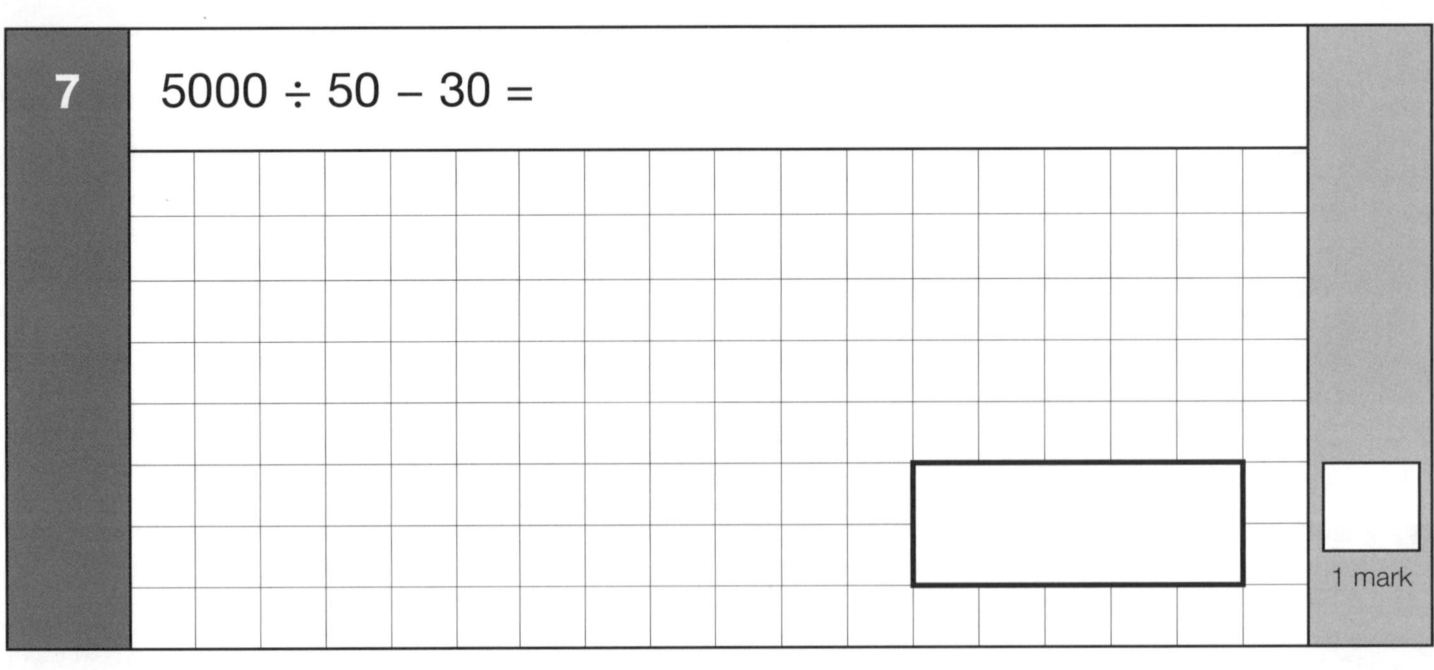
7
$5000 \div 50 - 30 =$
1 mark

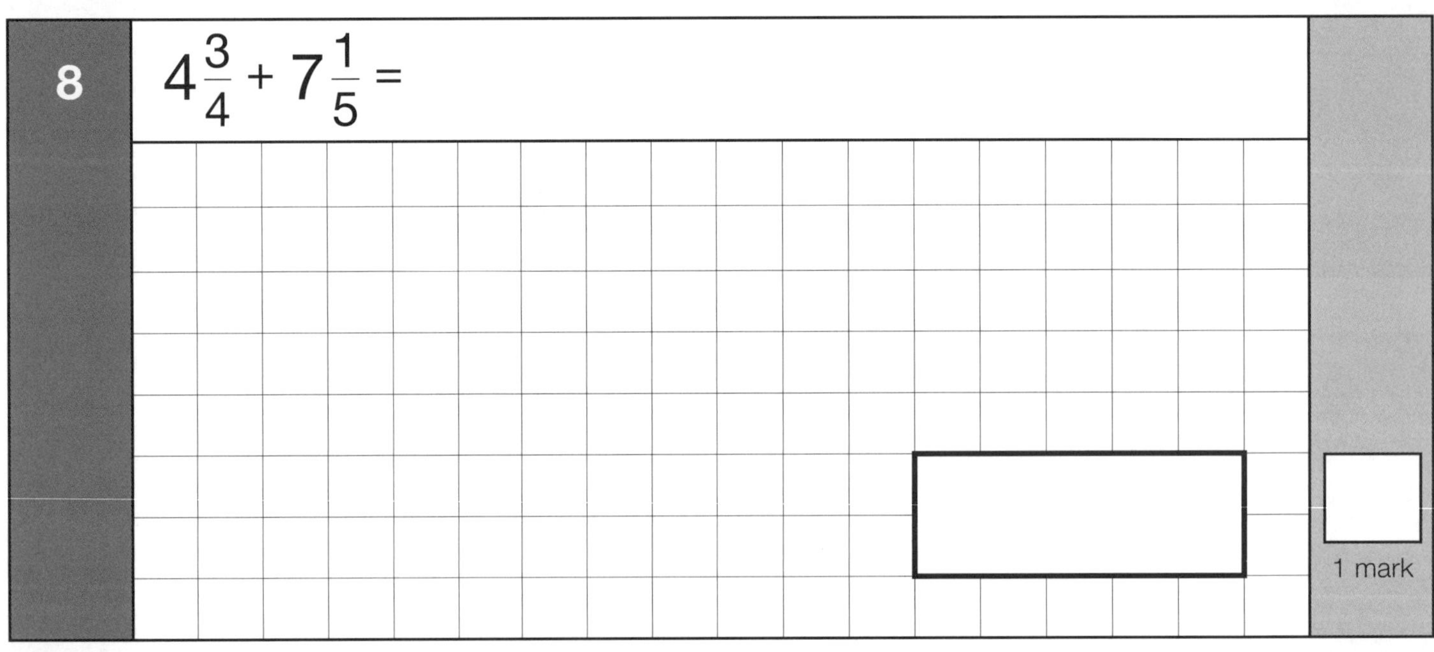
8
$4\frac{3}{4} + 7\frac{1}{5} =$
1 mark

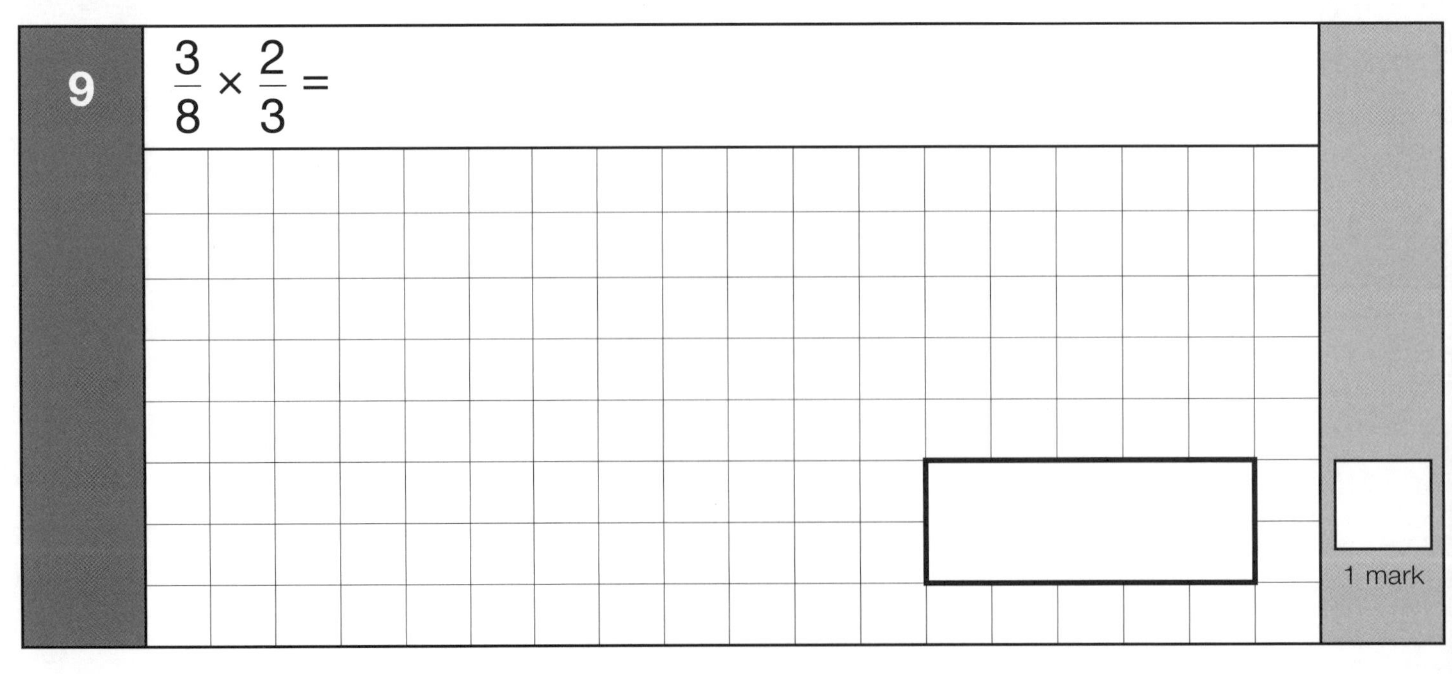
9
$\frac{3}{8} \times \frac{2}{3} =$
1 mark

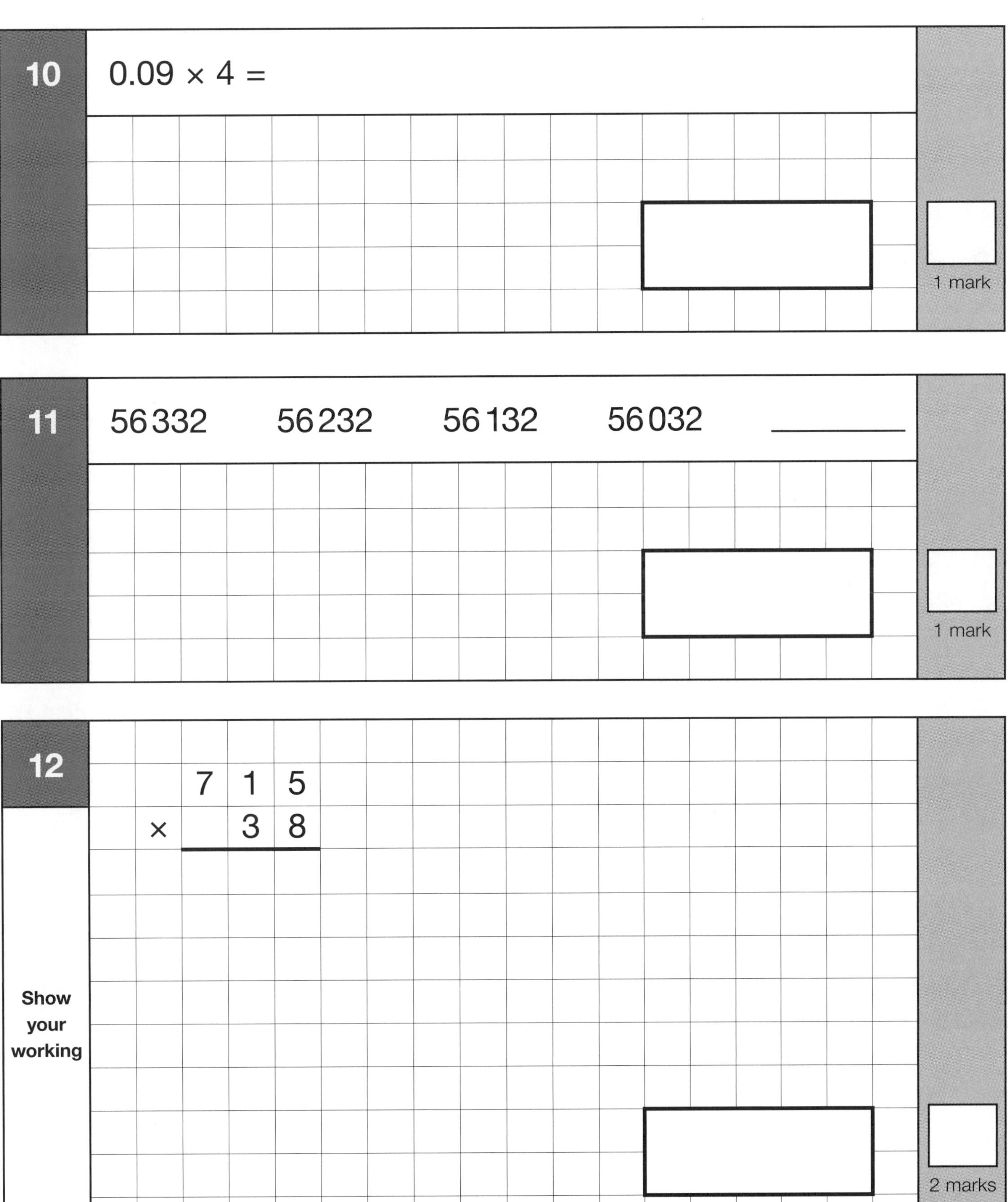
10
0.09 × 4 =
1 mark
11
56 332
56 232
56 132
56 032
1 mark
12
7 1 5
× 3 8
Show your working
2 marks

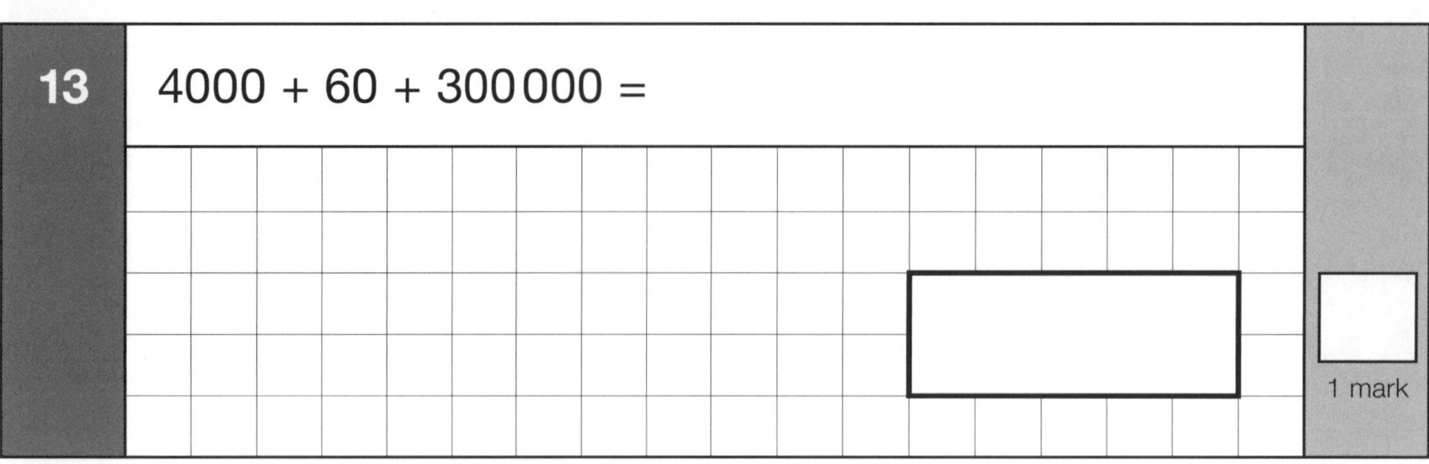

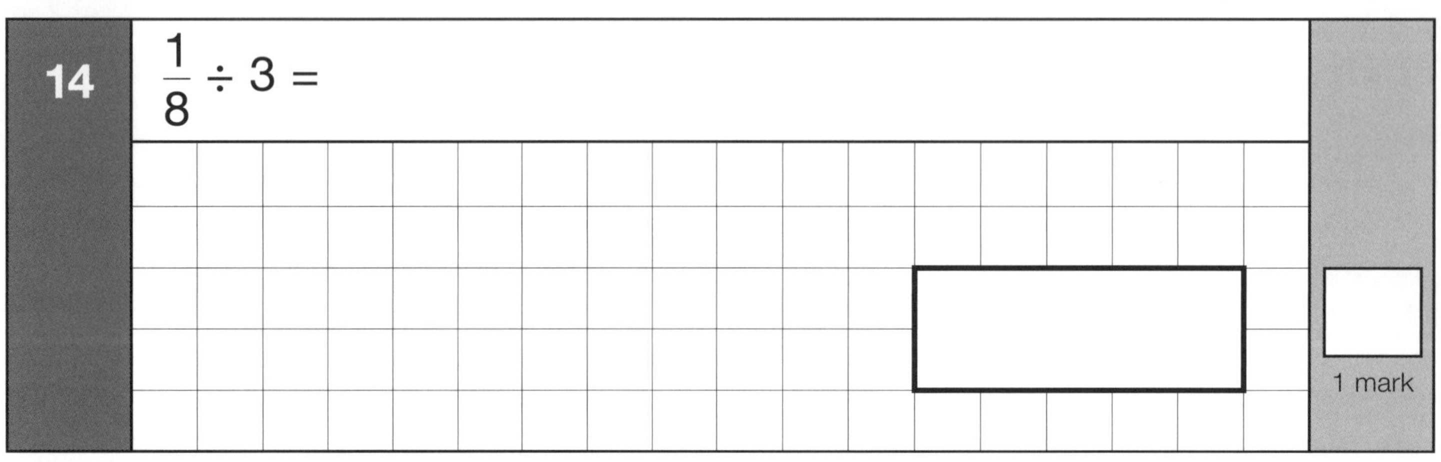

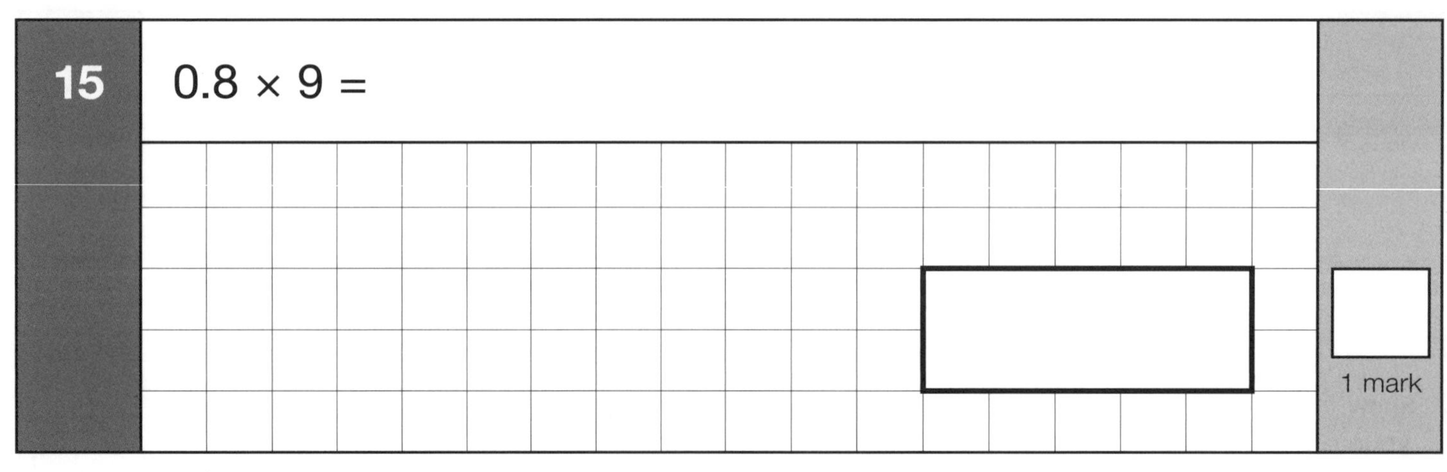

Total ________ / 17 marks

Arithmetic

TEST
5

- You have 20 minutes to complete this test.
- Calculator not allowed.
- Use the spaces provided for your workings. Where two marks are available, you may be awarded a mark for your workings.

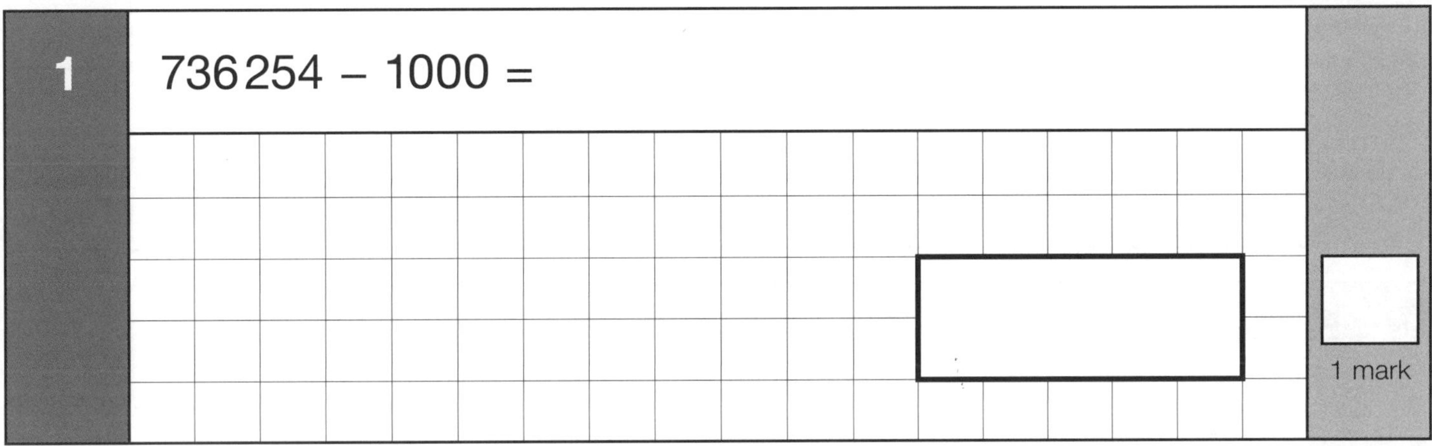

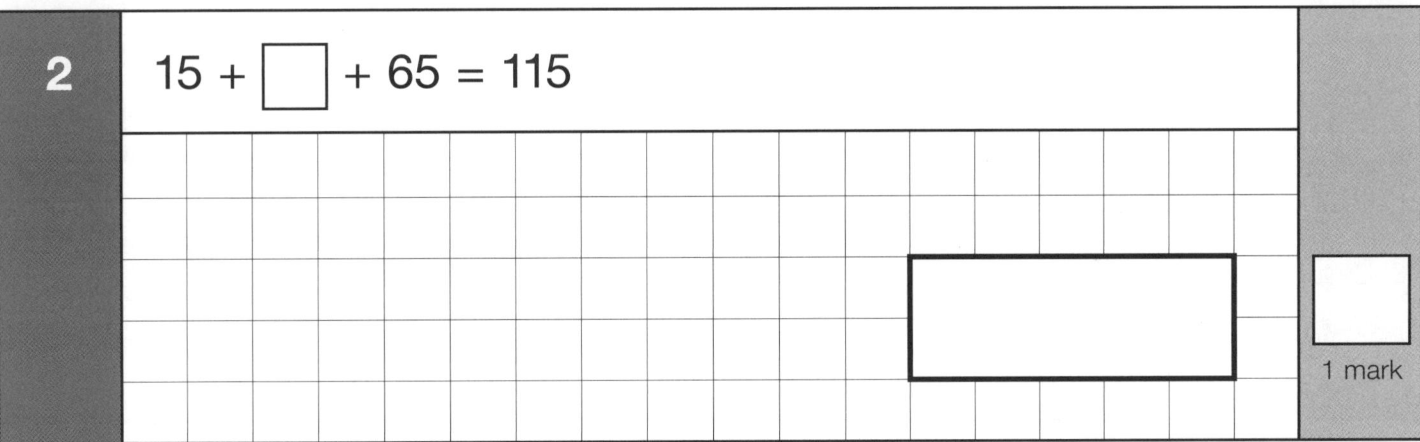

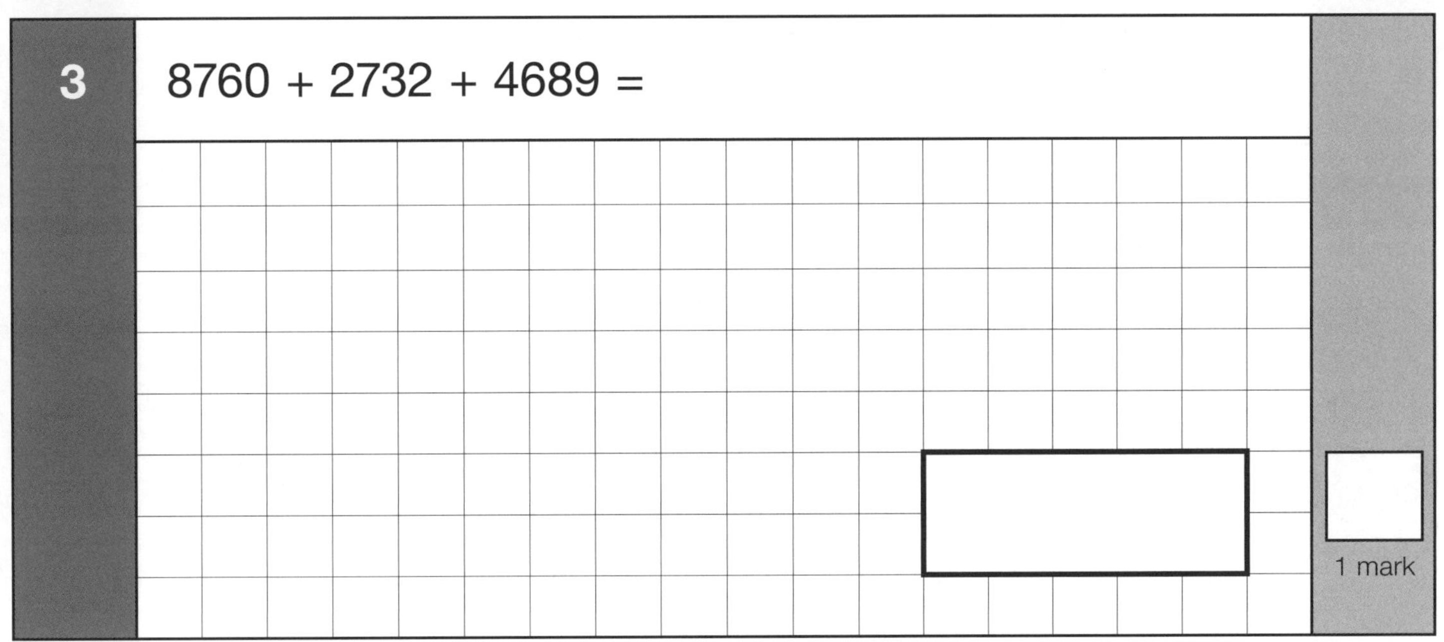

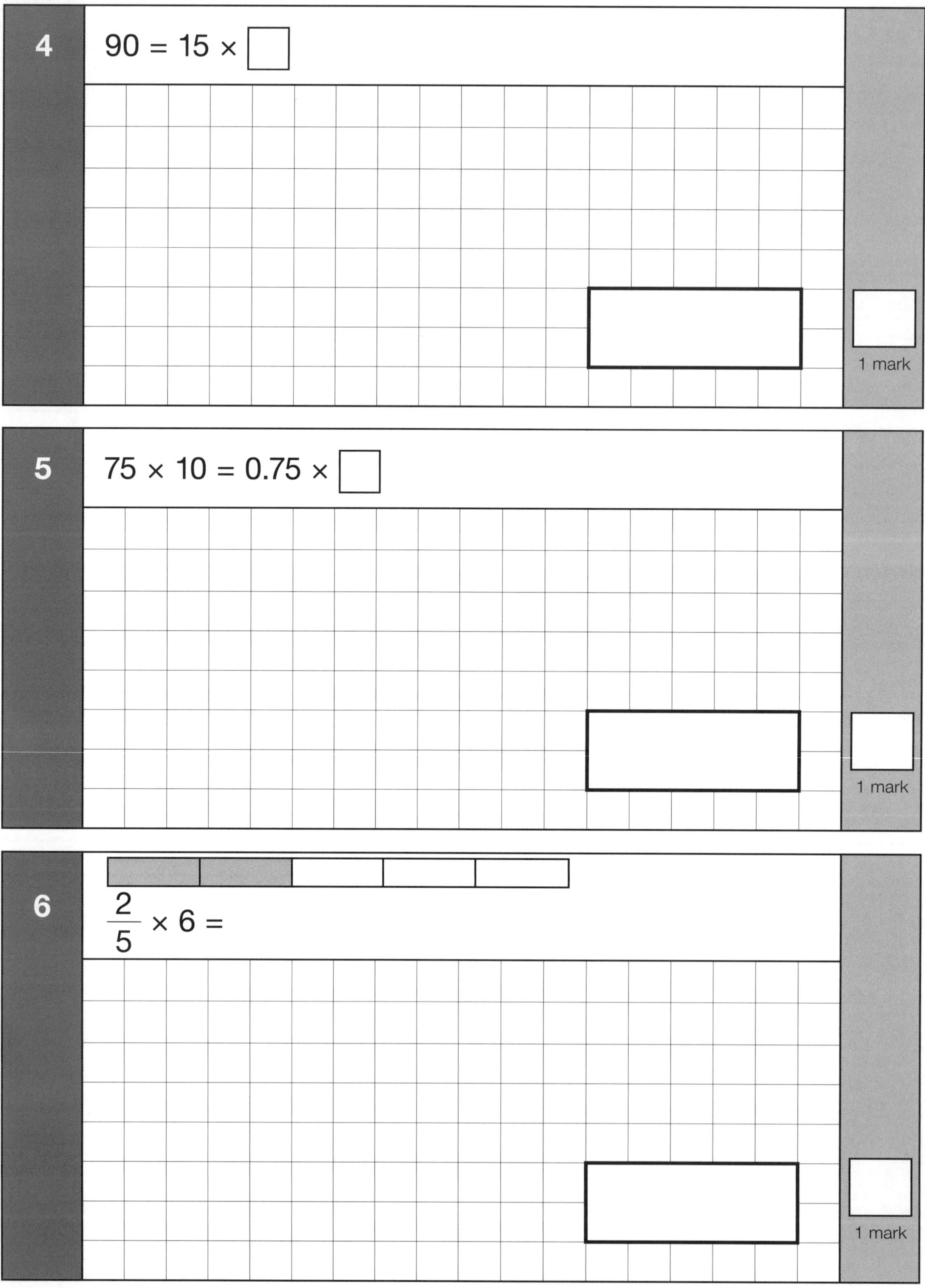

4
90 = 15 × ☐
1 mark
5
75 × 10 = 0.75 × ☐
1 mark
6
2/5 × 6 =
1 mark

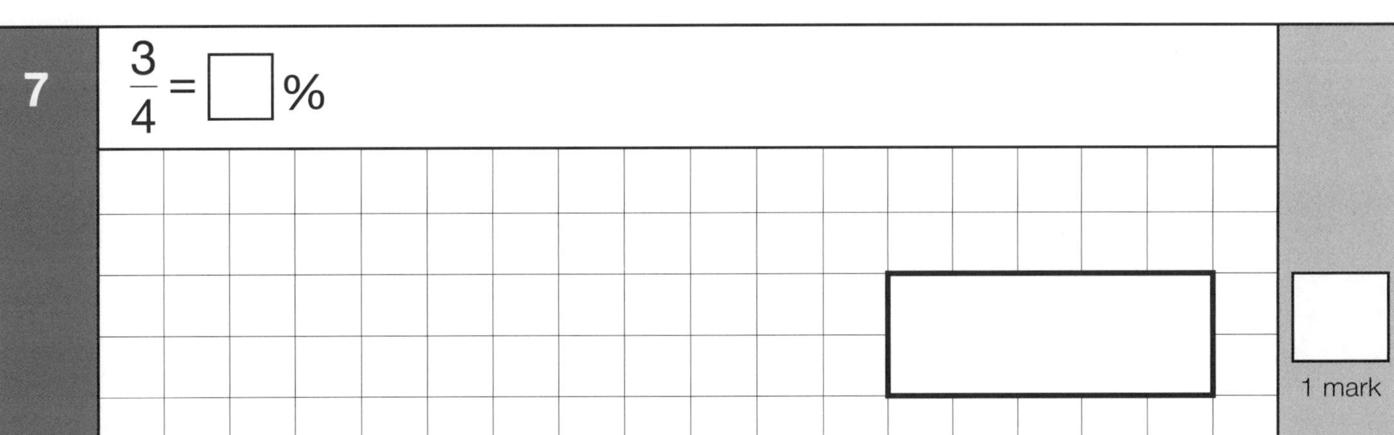
7
$\frac{3}{4} = \square\%$
1 mark

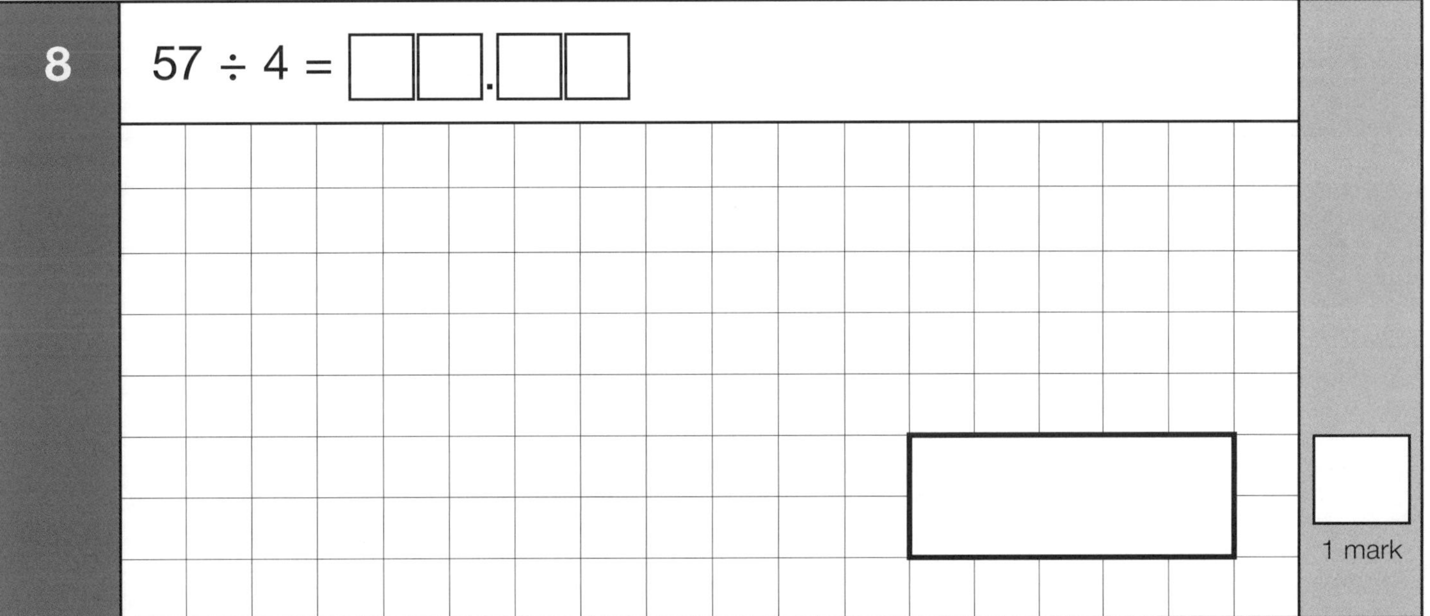
8
$57 \div 4 = \square\square.\square\square$
1 mark

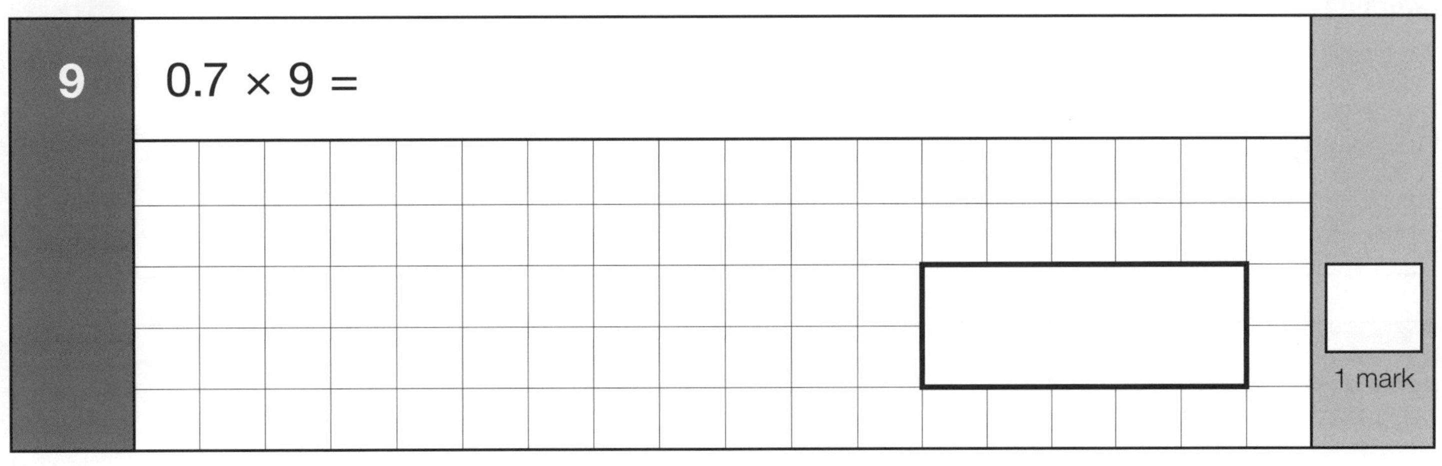
9
$0.7 \times 9 =$
1 mark

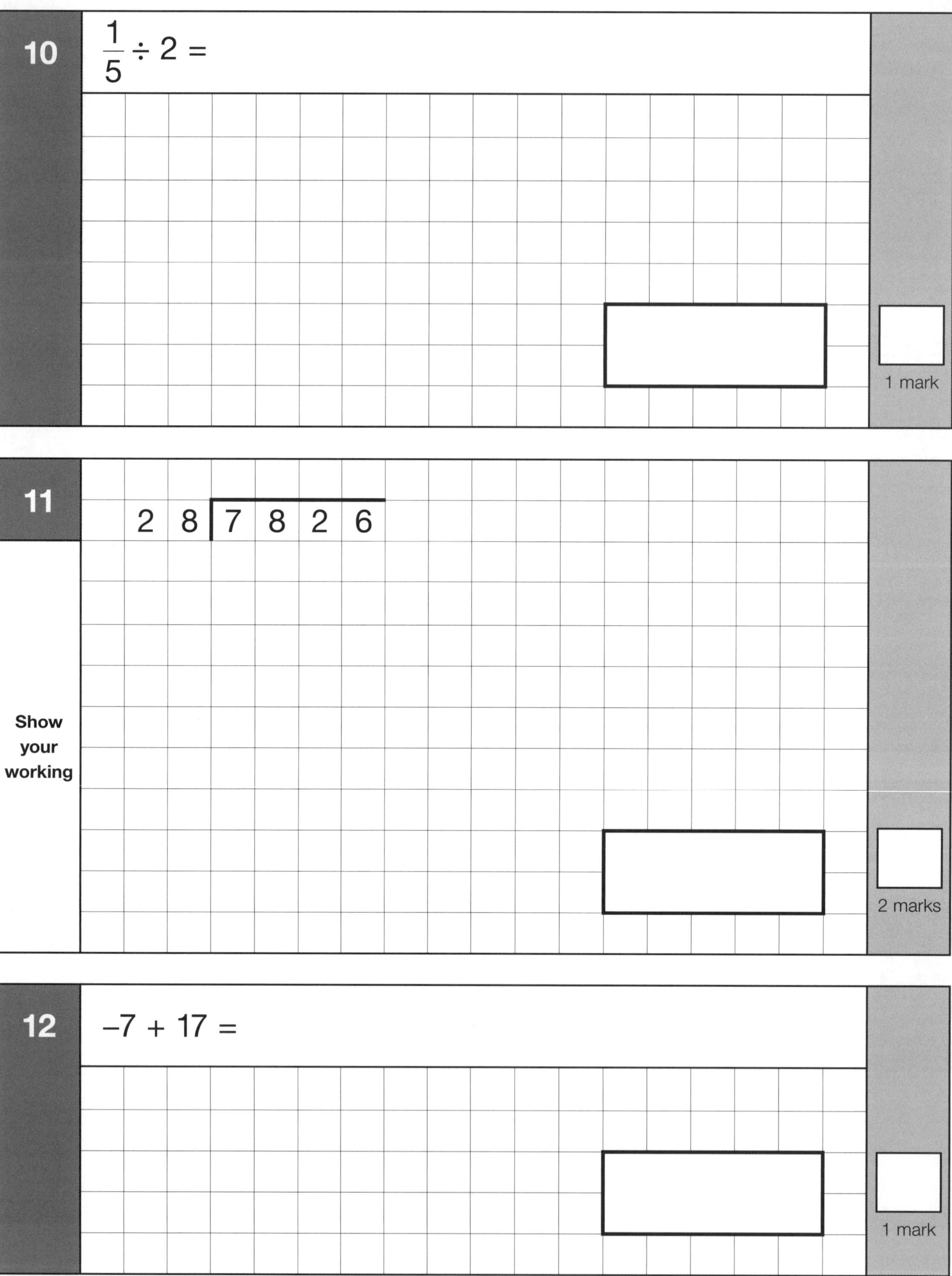
10
$\frac{1}{5} \div 2 =$
1 mark
11
2 8 | 7 8 2 6
Show your working
2 marks
12
$-7 + 17 =$
1 mark

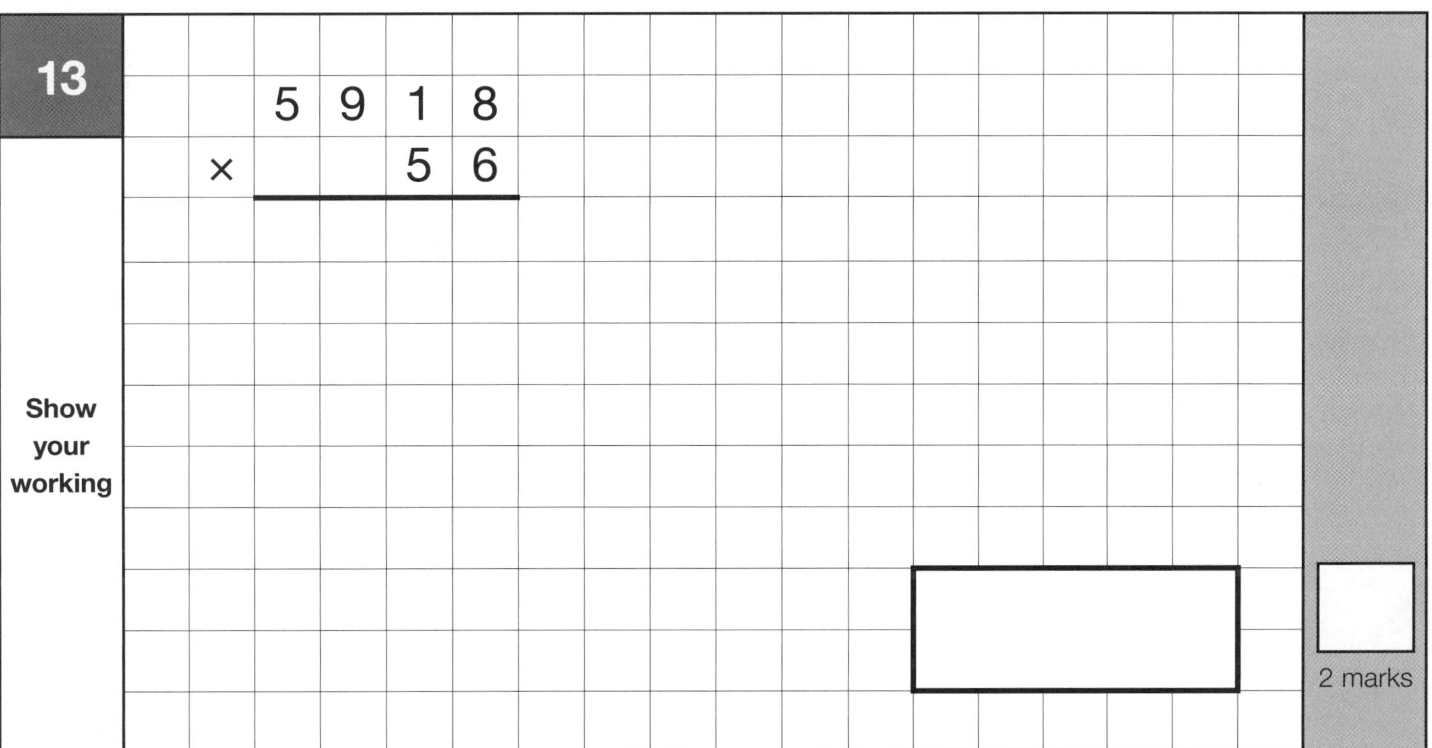

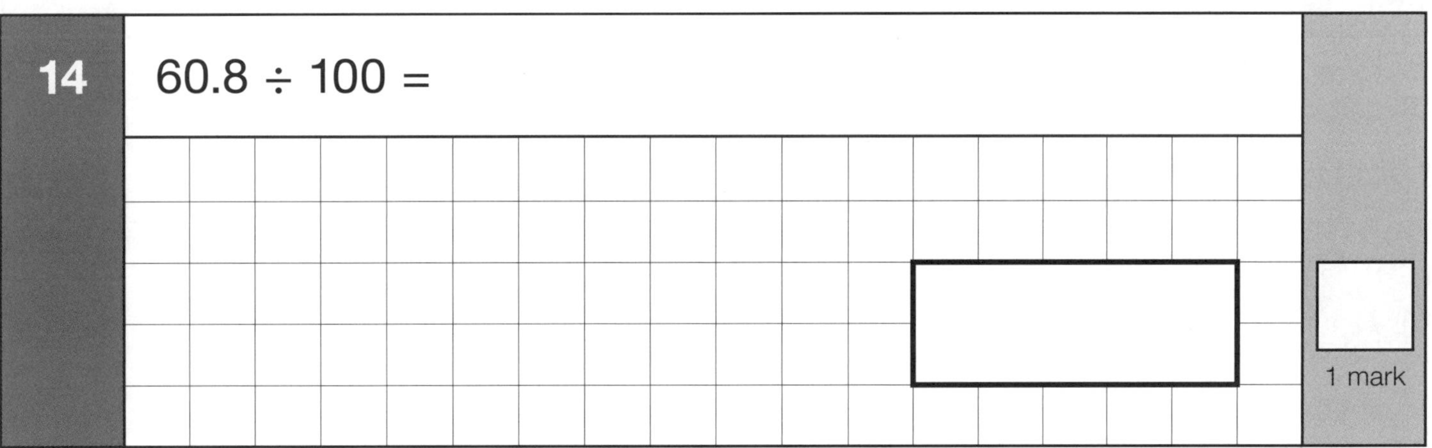

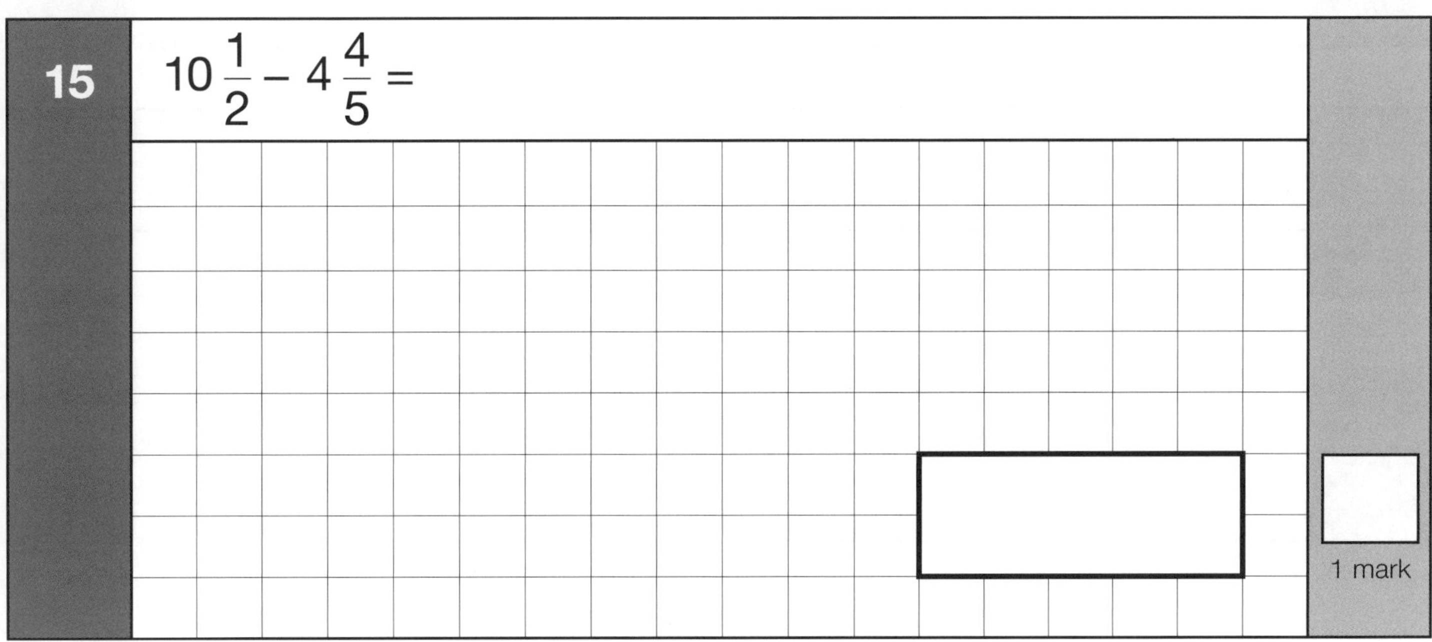

Total ________ / 17 marks

TEST 6

Arithmetic

- You have 20 minutes to complete this test.
- Calculator not allowed.
- Use the spaces provided for your workings. Where two marks are available, you may be awarded a mark for your workings.

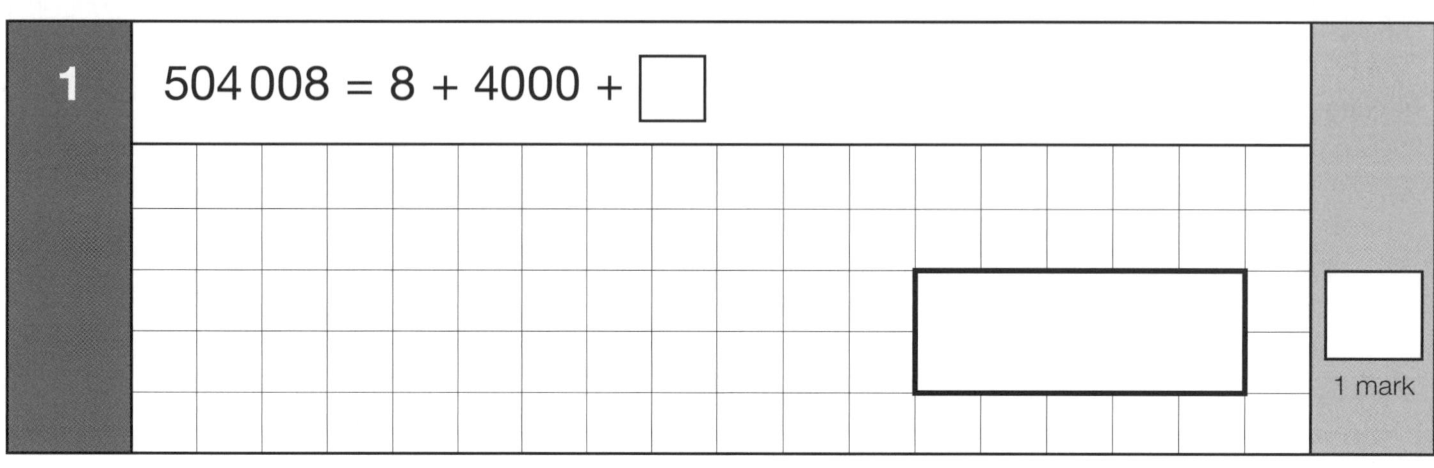

1 $504\,008 = 8 + 4000 + \square$

1 mark

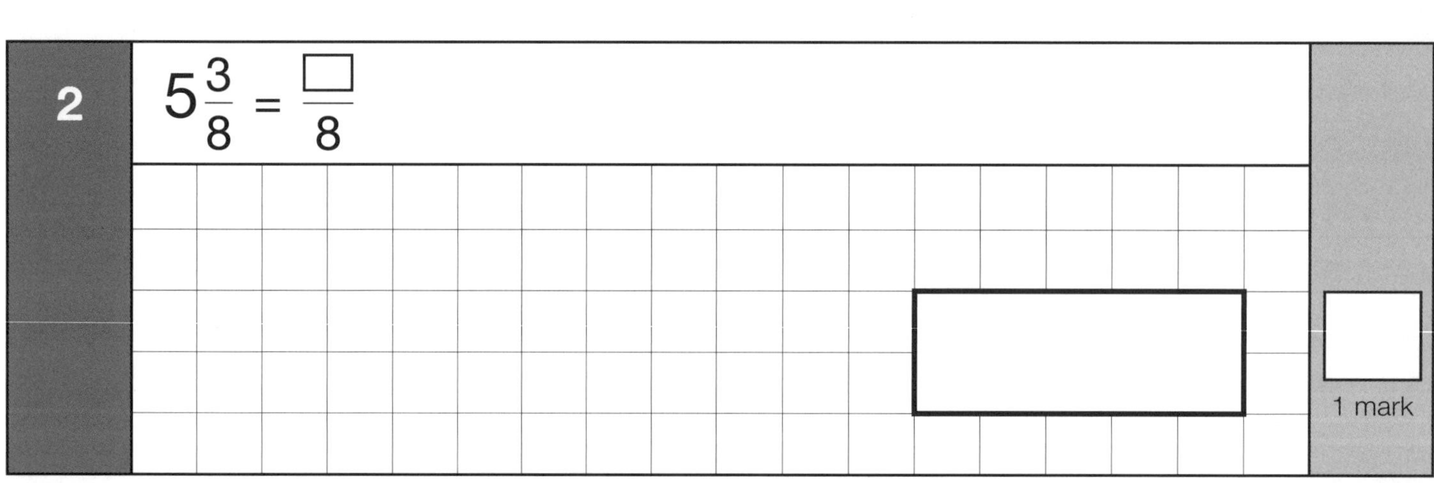

2 $5\frac{3}{8} = \frac{\square}{8}$

1 mark

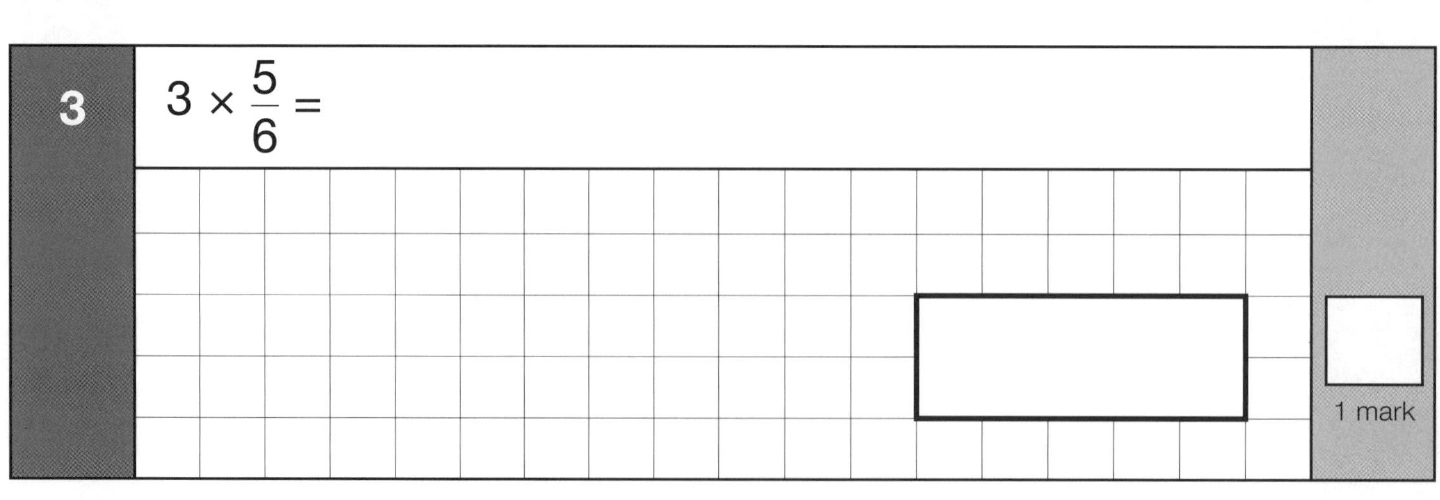

3 $3 \times \frac{5}{6} =$

1 mark

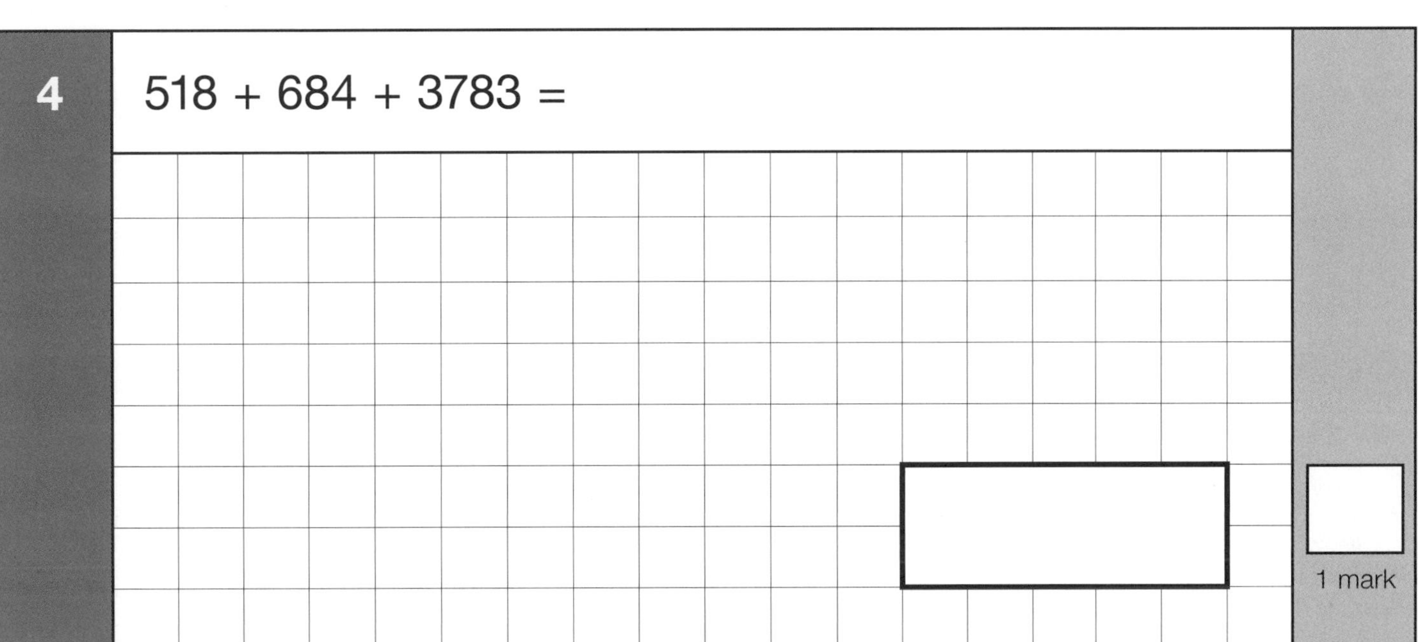

4 $518 + 684 + 3783 =$

1 mark

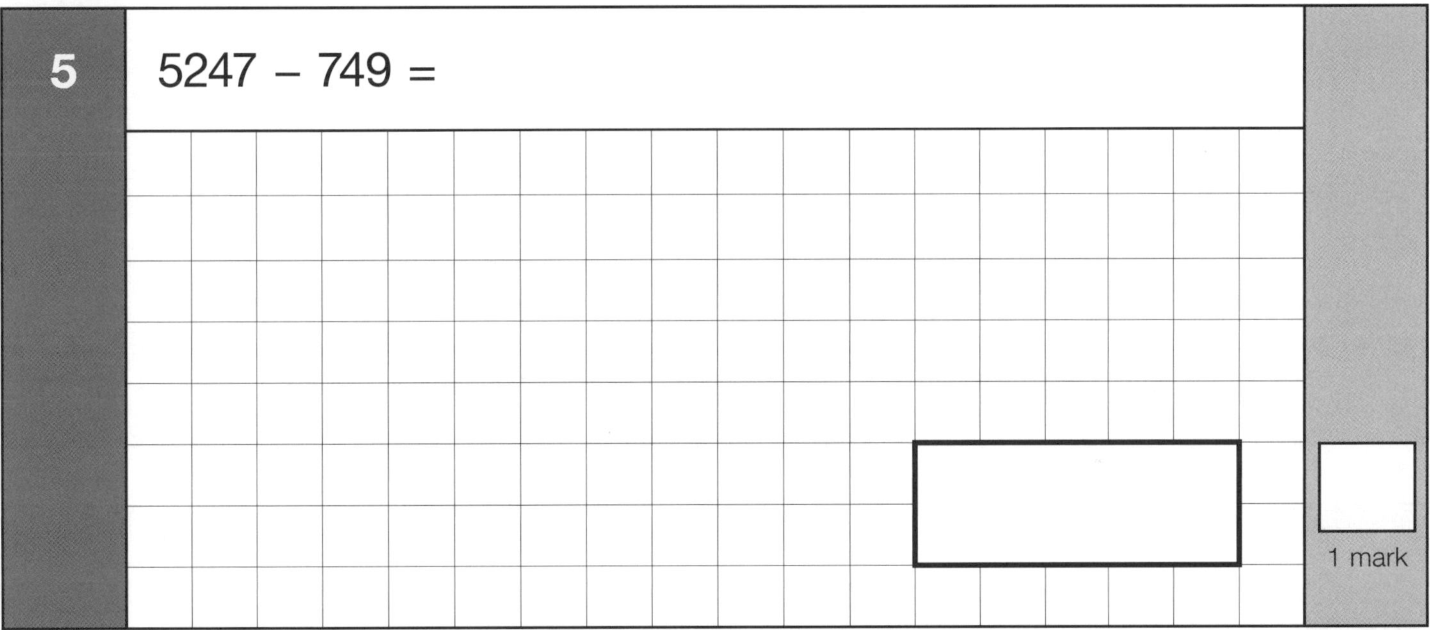

5 $5247 - 749 =$

1 mark

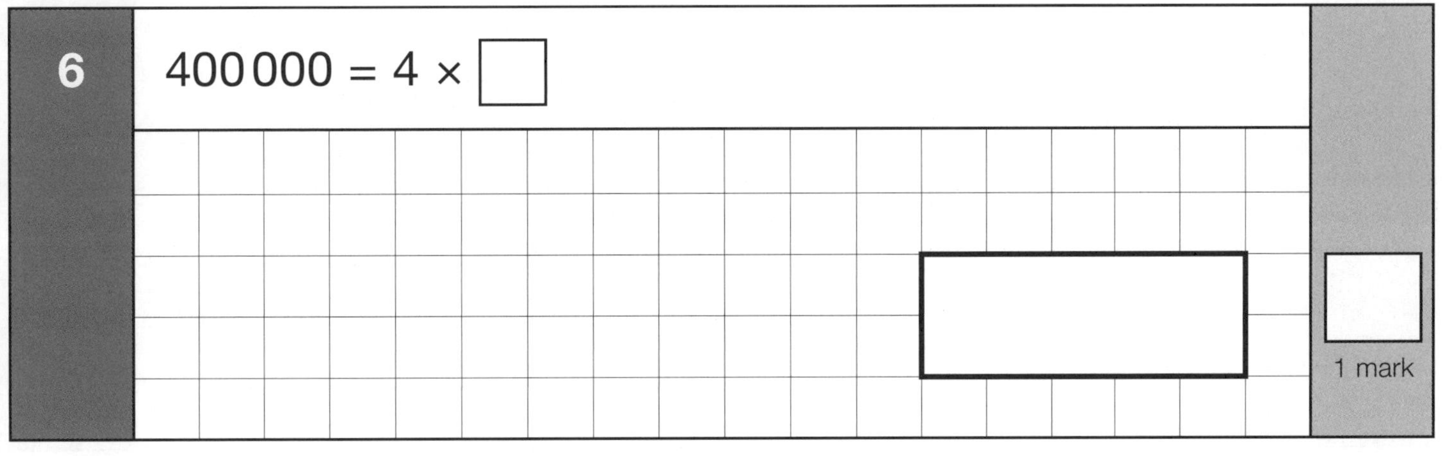

6 $400\,000 = 4 \times \square$

1 mark

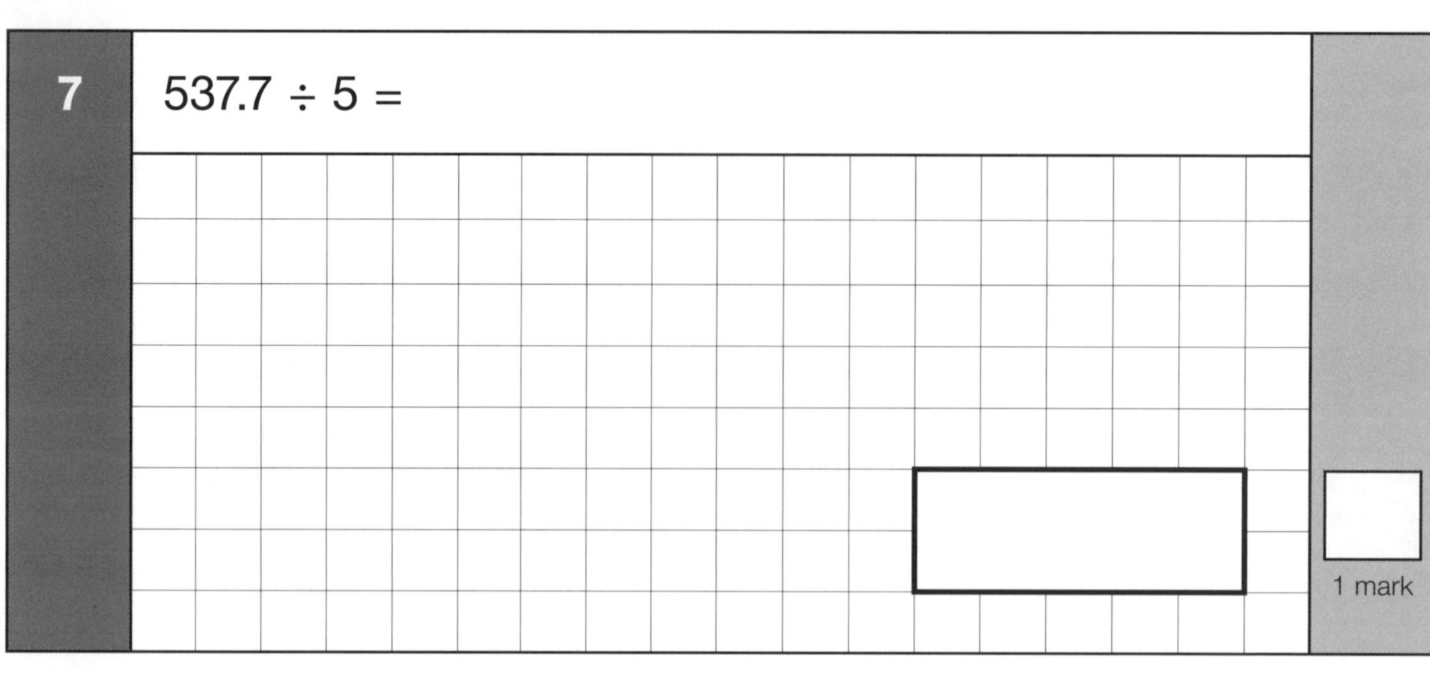
7
$537.7 \div 5 =$
1 mark

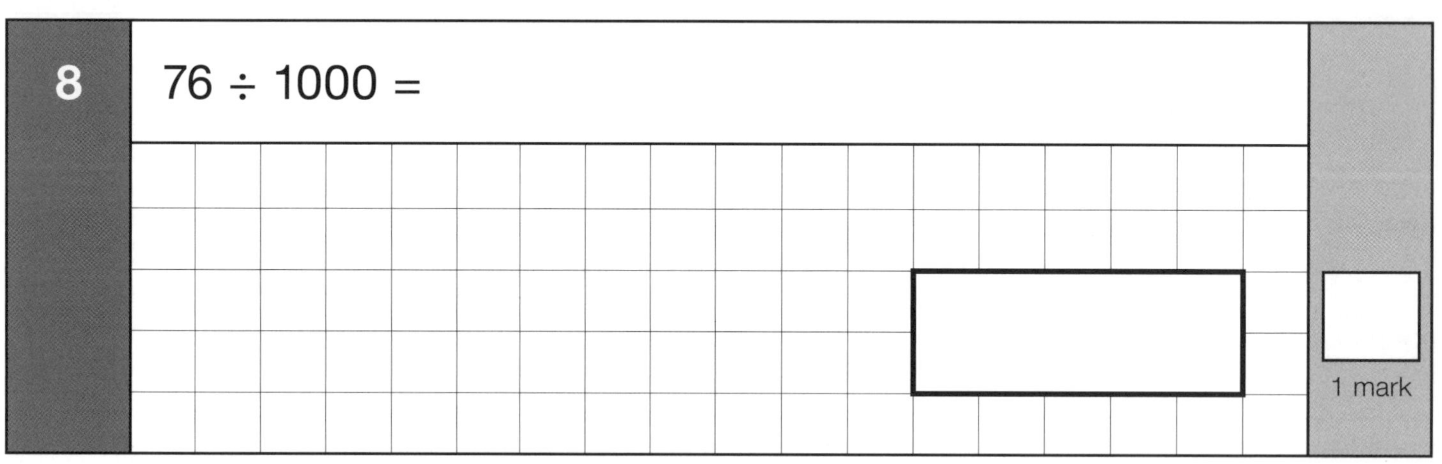
8
$76 \div 1000 =$
1 mark

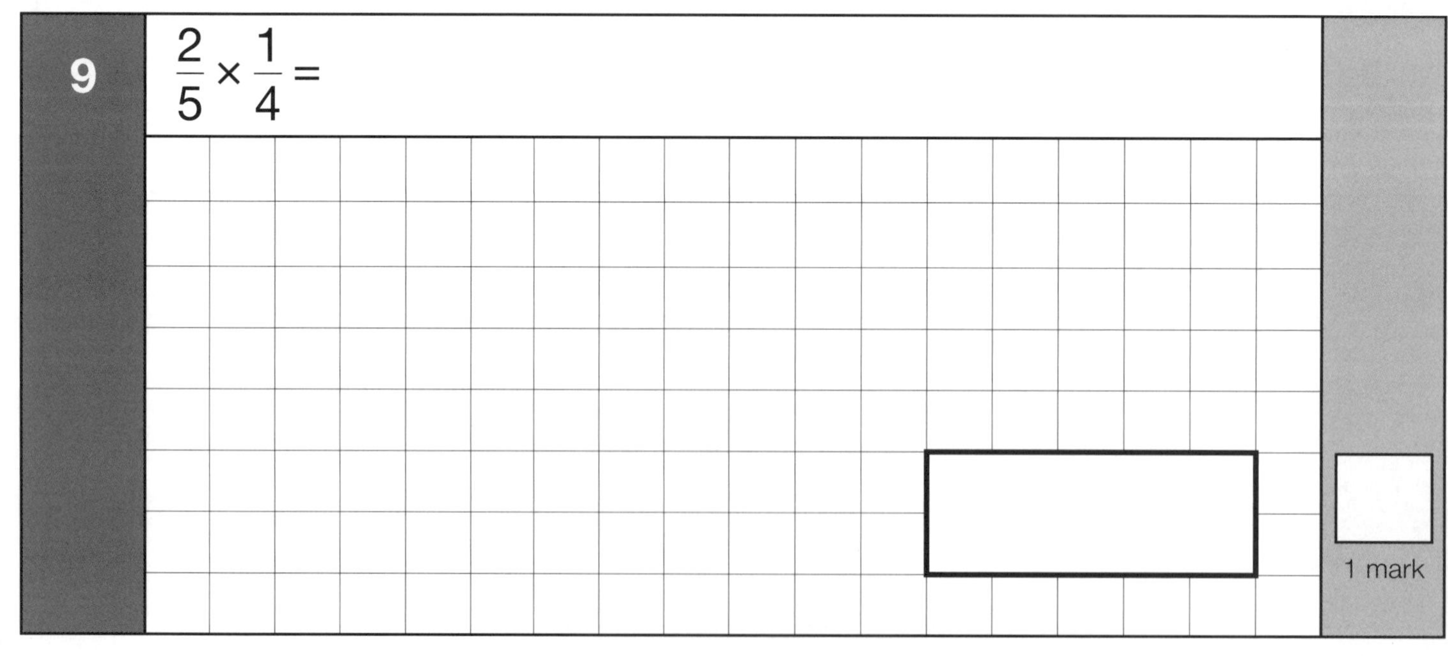
9
$\frac{2}{5} \times \frac{1}{4} =$
1 mark

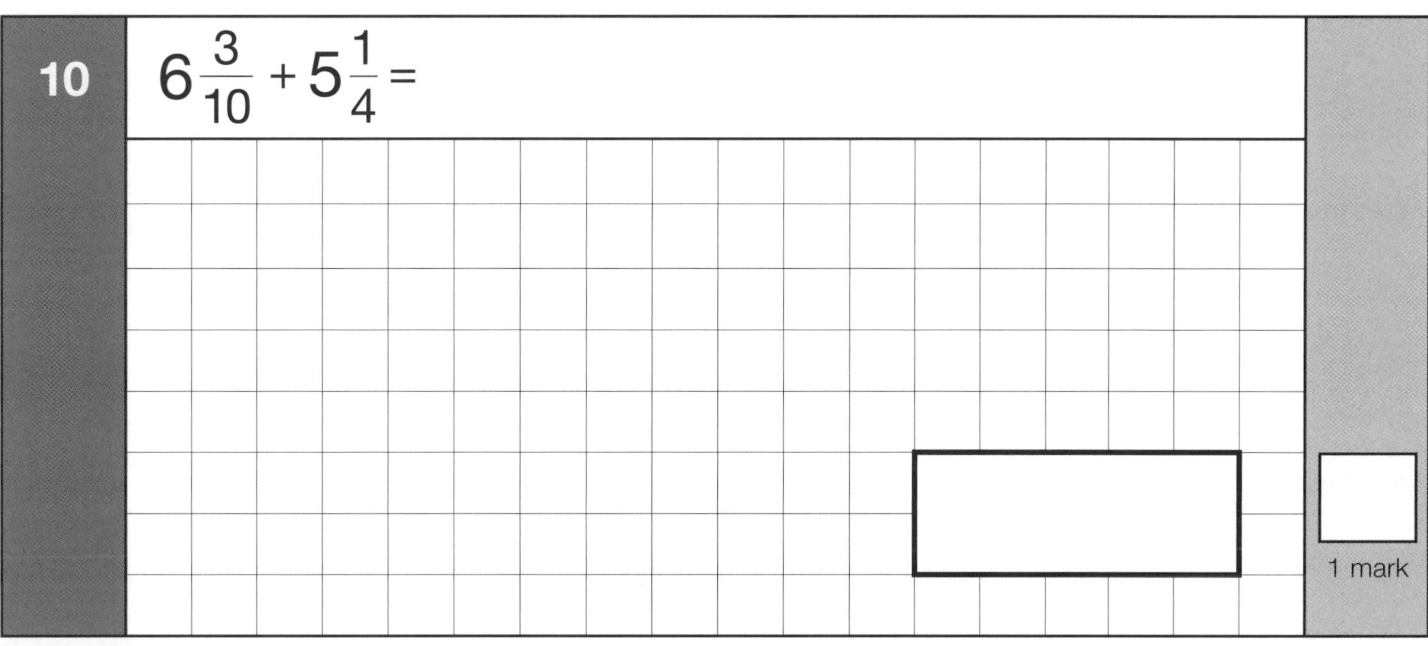

10
$6\frac{3}{10} + 5\frac{1}{4} =$
1 mark

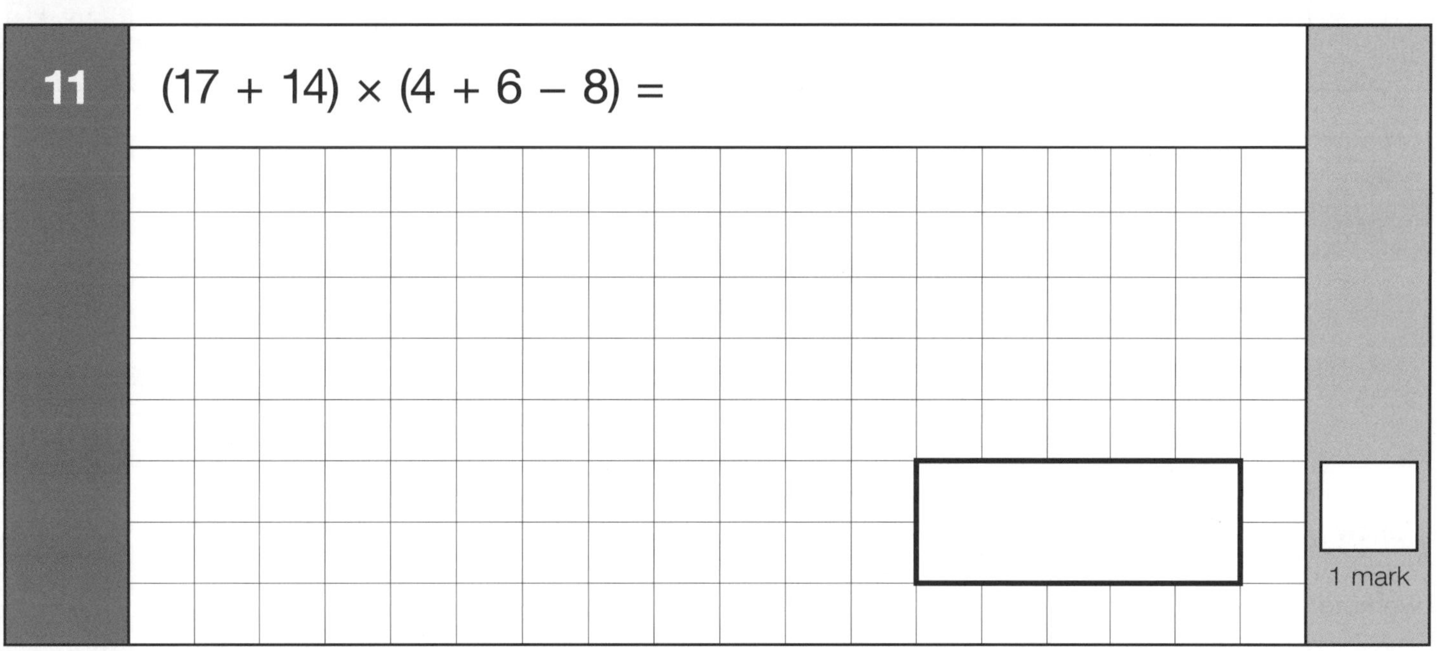

11
$(17 + 14) \times (4 + 6 - 8) =$
1 mark

12

$$\begin{array}{r} 7593 \\ \times \quad 74 \\ \hline \end{array}$$

Show your working

2 marks

13

$$24 \overline{)2082}$$

Show your working

2 marks

Total ________ / 15 marks

Using Mathematics

TEST 1

- You have 20 minutes to complete this test.
- Calculator not allowed.

1 Which number comes next in this sequence?

48 56 64 72

1 mark

2

What time is shown on this clock?

1 mark

3 5727 5277 5272 5772

Write these numbers in order, smallest first.

smallest

1 mark

4 Part of this shape is missing.

The dotted line is a line of symmetry.

Complete the shape.

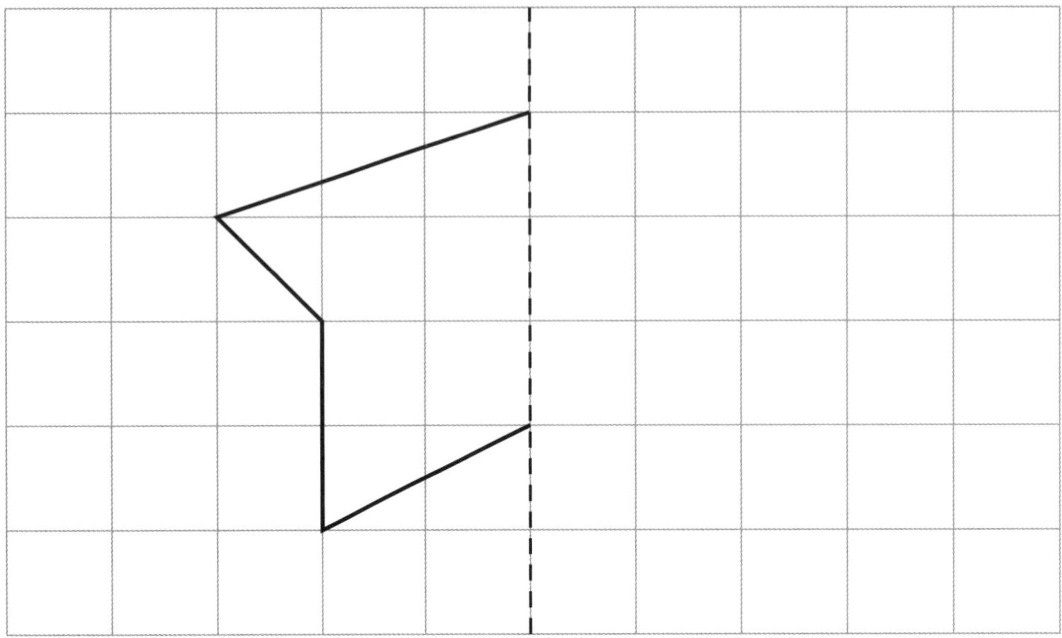

1 mark

5 Tom has 217 minutes left on his phone.

He uses 83 minutes.

He gets another 350 minutes.

How many minutes does Tom have on his phone?

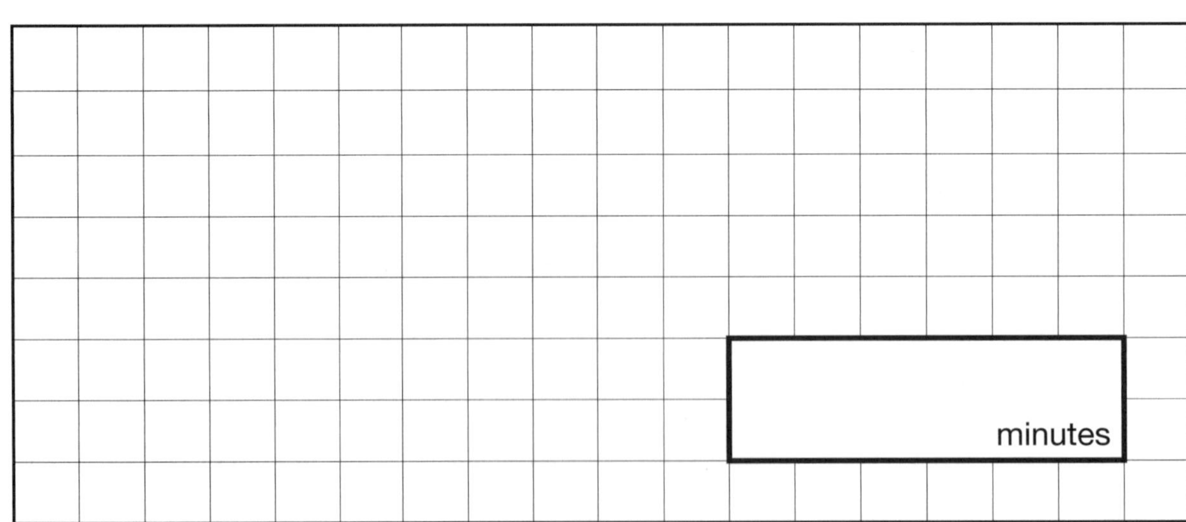

1 mark

6 Four hundred and sixty thousand, three hundred and five.

Write this number in digits.

1 mark

7 Two prime numbers total 31

What are the two numbers?

1 mark

8 **Write the equivalent fractions shown by the shading in these shapes.**

2 marks

9 This is a train timetable.

Little Oak	10.30	11.30	12.30	13.30
Elmstree		12.10		
Ashton	11.25		13.25	14.25
Beechwood			13.43	14.41
Birchly	12.03	12.58	14.05	15.06

Jack arrives at Little Oak at 11.00

He catches the next train to Ashton.

When will he arrive at Ashton?

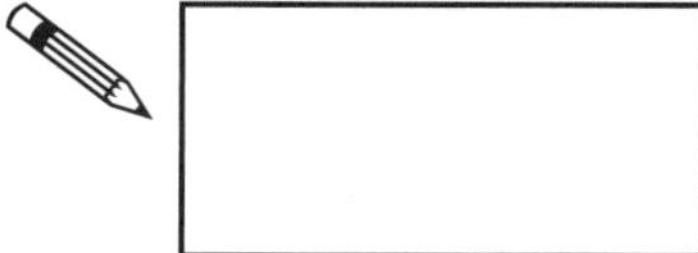

1 mark

10 What is 25% of 600?

1 mark

11 Aisha adds two numbers; the total is 15

The difference between the two numbers is 5

What are the two numbers?

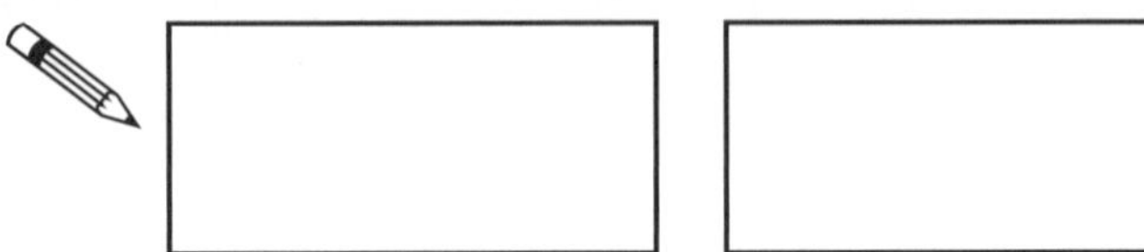

1 mark

12 This is a cube.

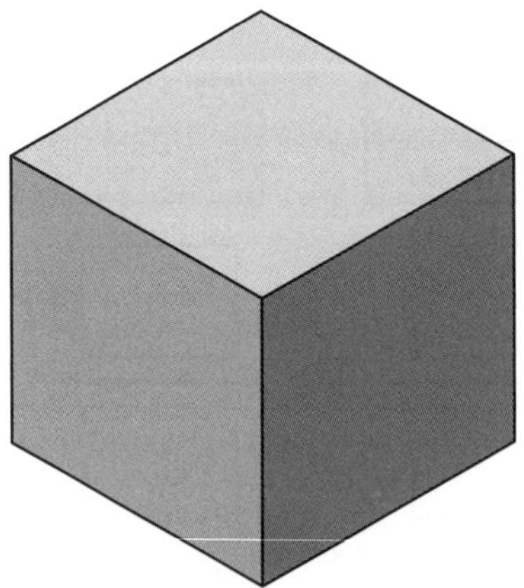

Complete these sentences.

A cube has [ ] faces.

A cube has [ ] edges.

A cube has [ ] vertices.

3 marks

Total ________ / 15 marks

Using Mathematics

TEST 2

- You have 20 minutes to complete this test.
- Calculator not allowed.

1 Which three-digit number can you make from these three cards?

3 units	4 hundreds	9 tens

1 mark

2 Tick (✔) the right angles in this shape.

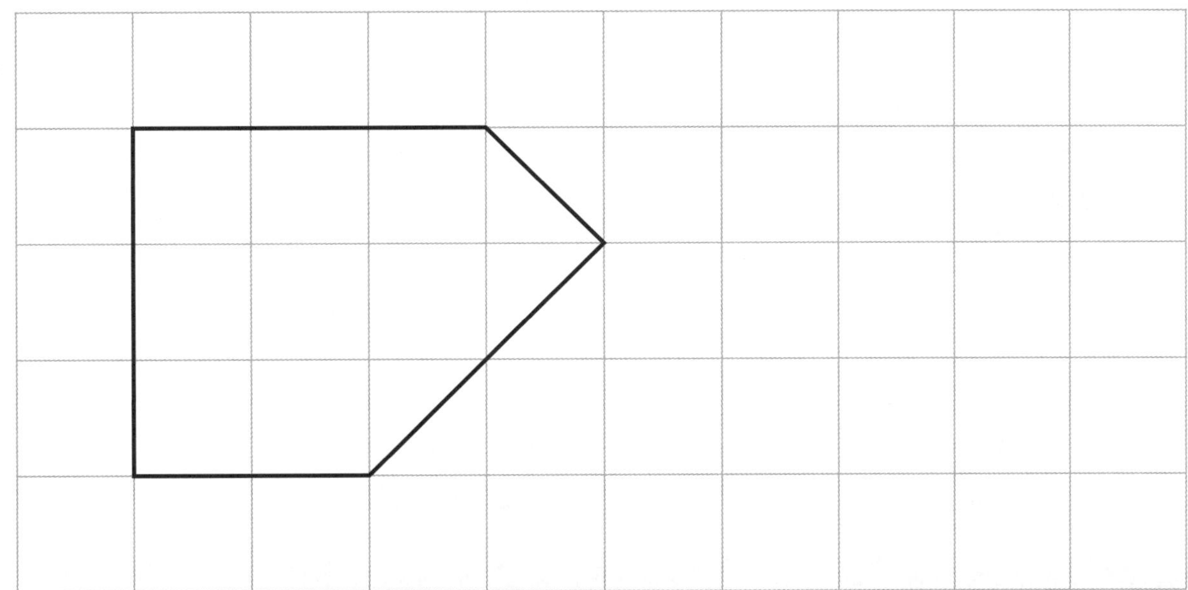

2 marks

3 Round 5199 to the nearest thousand.

1 mark

4 Ben sees MCMLXXX written on a building.

MCMLXXX stands for a year.

Name the year.

1 mark

5 **Circle the factors of 30.**

3 4 6 15 60 150

1 mark

6 **Write these fractions in order, starting with the largest.**

$\frac{3}{5}$ $\frac{7}{10}$ $\frac{61}{100}$ $\frac{13}{20}$

largest

1 mark

7

Shape A

Shape B

Explain why Shape A is a regular pentagon and Shape B is not.

1 mark

8 1 inch is about 2.5 centimetres.

How many centimetres is 12 inches?

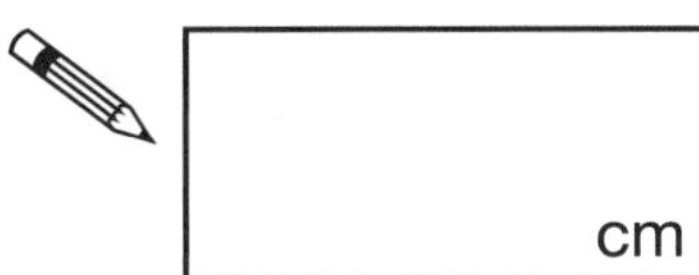

1 mark

9 The temperature outside a greenhouse is –4°C.

The temperature inside the greenhouse is 4°C.

What is the difference between the two temperatures?

1 mark

10 William sees a special offer in a shop.

He can buy four cans of soup for £1.50

How much would William pay for twelve cans of soup?

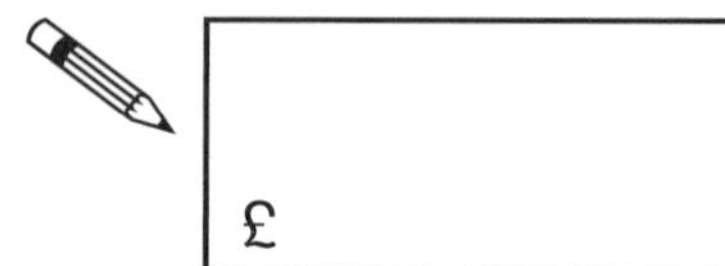

£

1 mark

11 Here are three number cards:

7	1	6

Nishi arranges the cards to make three-digit numbers.

List all the odd numbers that Nishi can make.

2 marks

12 Find the missing number.

$0.7 \times \square = 4.2$

1 mark

13 This pie chart shows the sports chosen by 80 children.

Here are some facts about the pie chart:

- 30 children chose tennis.
- $\frac{1}{4}$ of the children chose football.
- The same number of children chose gymnastics as chose rugby.

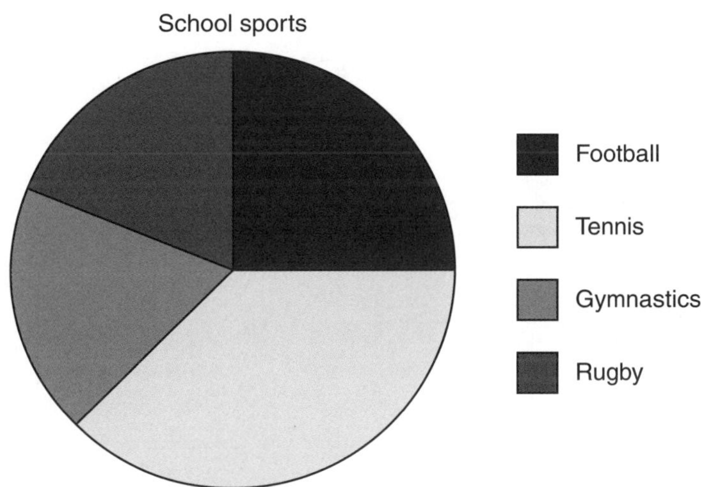

How many children chose rugby?

Show your working

children

2 marks

Total ________ / 16 marks

TEST 3

Using Mathematics

- You have 20 minutes to complete this test.
- Calculator not allowed.

1 This bag has 5 black balls and 3 white balls.

What fraction of the balls is black?

1 mark

2 A train has 8 coaches.

Each coach has 72 seats.

The ticket collector says, 'I know $70 \times 8 = 560$'

What must he add to 560 to find how many seats there are in the train altogether?

1 mark

3 **a.** Round 5.17 to:

i. one decimal place. **ii.** the nearest whole number.

1 mark

b. Round 14.73 to:

i. one decimal place. **ii.** the nearest whole number.

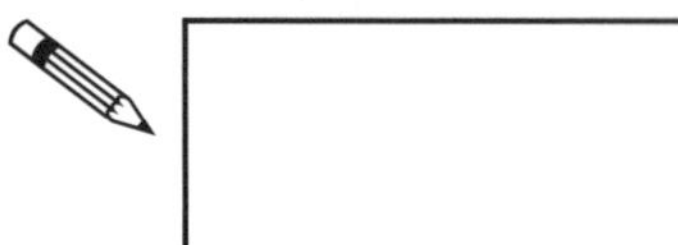

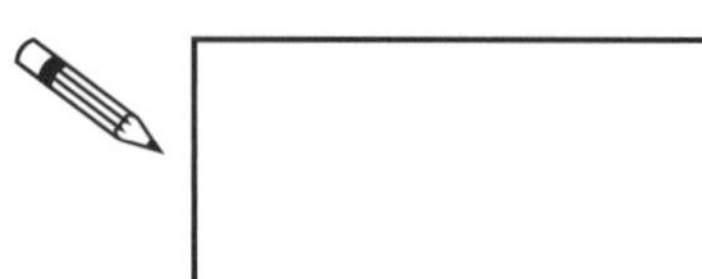

1 mark

4 Draw lines to match the same lengths.

One has been drawn for you.

4.6m	4.6cm
46mm	0.46km
460m	460cm
4.6mm ———	0.46cm
46cm	0.46m

2 marks

5 Polly calculates 87 × 73

Round each number to the nearest ten.

Use the rounded numbers to give an estimated answer to Polly's calculation.

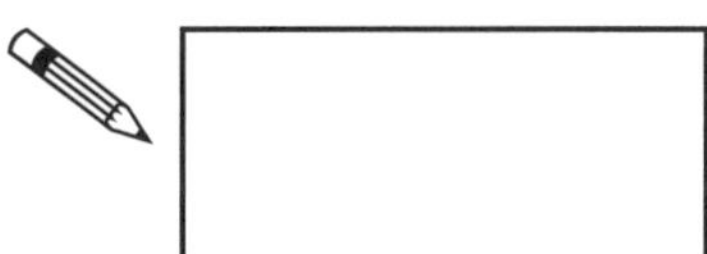

1 mark

6 What number comes next in this sequence?

536 280 537 280 538 280 539 280

1 mark

7

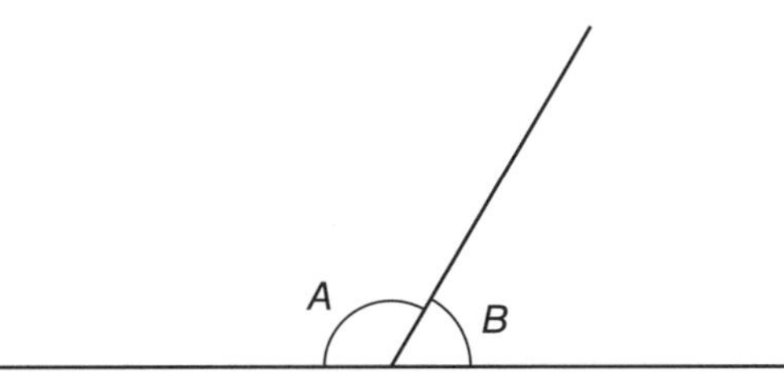

What is the total of angle A and angle B?

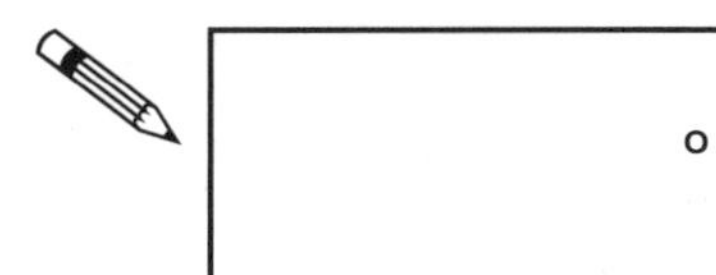

1 mark

8 In 2 hours a machine makes 800 packs of biscuits.

64 packs of biscuits fill a box.

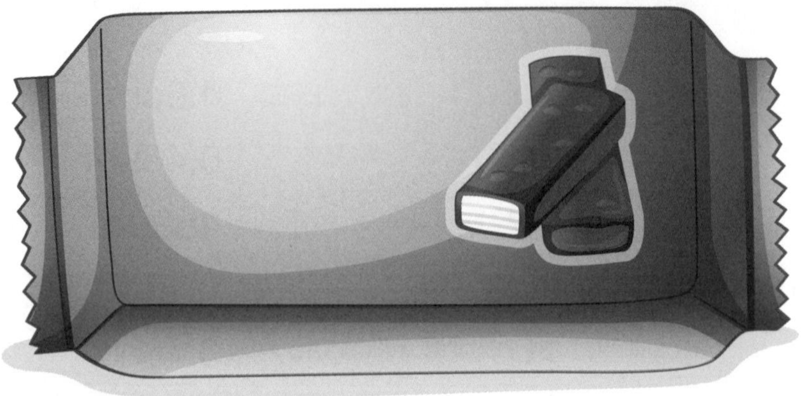

a. How many boxes will be filled completely in 2 hours?

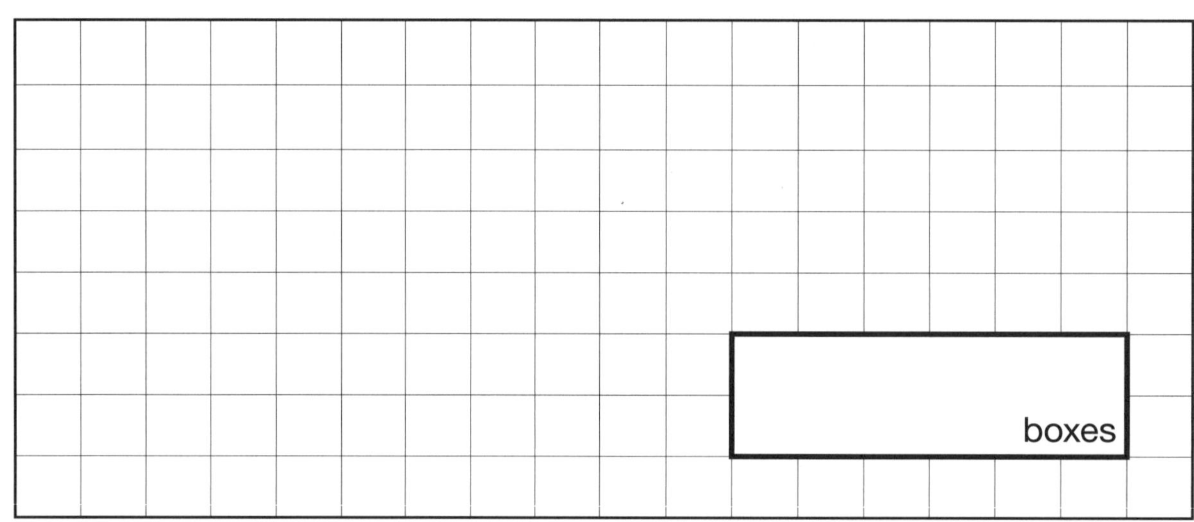

1 mark

b. How many packs of biscuits will be made in 12 hours?

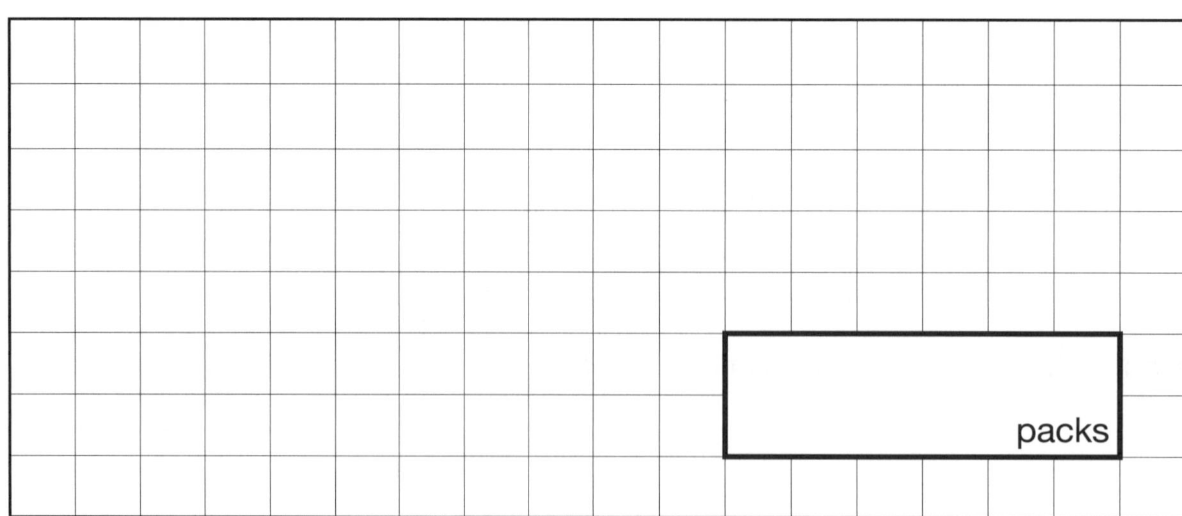

1 mark

9 Josef says, 'I am thinking of a number. I multiply the number by 6 and subtract 8'

Tick (✔) the expression that shows Josef's calculation algebraically.

- ☐ $6n + 8$
- ☐ $8 - 6n$
- ☐ $6(n - 8)$
- ☐ $6n - 8$
- ☐ $8 \times 6 - n$

1 mark

10

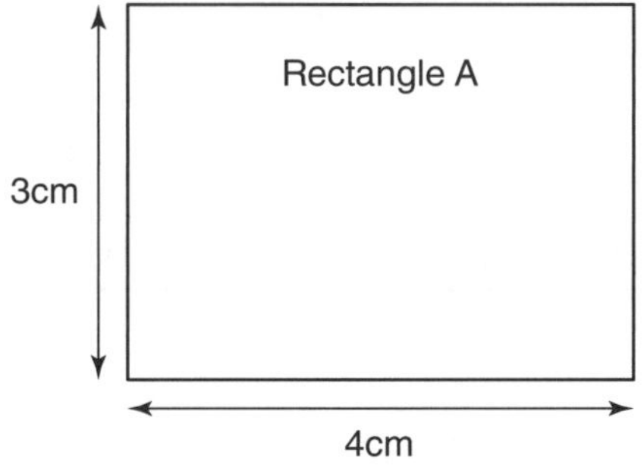

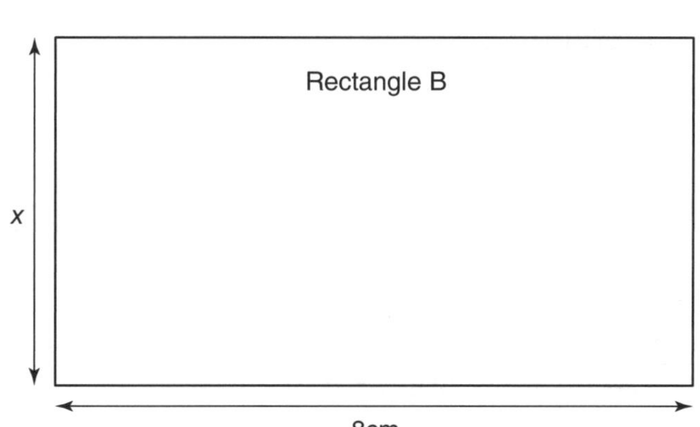

Rectangle A is enlarged and is drawn as rectangle B.

How long is side *x*?

1 mark

11

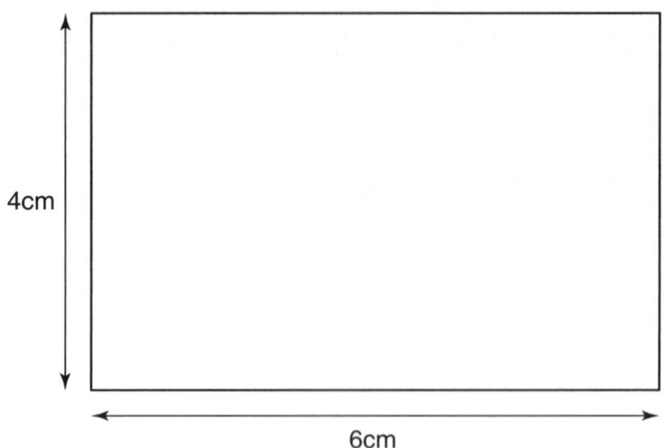

a. Give the length and width of a different rectangle with the same area.

length = ______ cm, width = ______ cm

1 mark

b. Give the length and width of a different rectangle with the same perimeter.

length = ______ cm, width = ______ cm

1 mark

12 Circle the fractions that will simply to $\frac{3}{5}$

$\frac{45}{75}$ $\frac{18}{30}$ $\frac{10}{15}$ $\frac{25}{40}$ $\frac{36}{60}$

2 marks

Total ______ / 17 marks

Using Mathematics

TEST 4

- You have 20 minutes to complete this test.
- Calculator not allowed.

1 This is a rectangle.

Tick (✔) **two** correct statements.

☐ The two bold lines are perpendicular.

☐ The bold and thin lines are perpendicular.

☐ The two bold lines are parallel.

☐ The bold and thin lines are parallel.

2 marks

2 Put these numbers in order, starting with the smallest.

39.39 9.93 3.09 9.09 3.39

smallest				

1 mark

3 Circle the **square** numbers.

1   5   36   46   64   91

2 marks

4 This table shows Obi's journey from York to Carlisle.

	Time taken (minutes)
York to Newcastle	55
Waiting time	35
Newcastle to Carlisle	50

How long did it take Obi to go from York to Carlisle in hours and minutes?

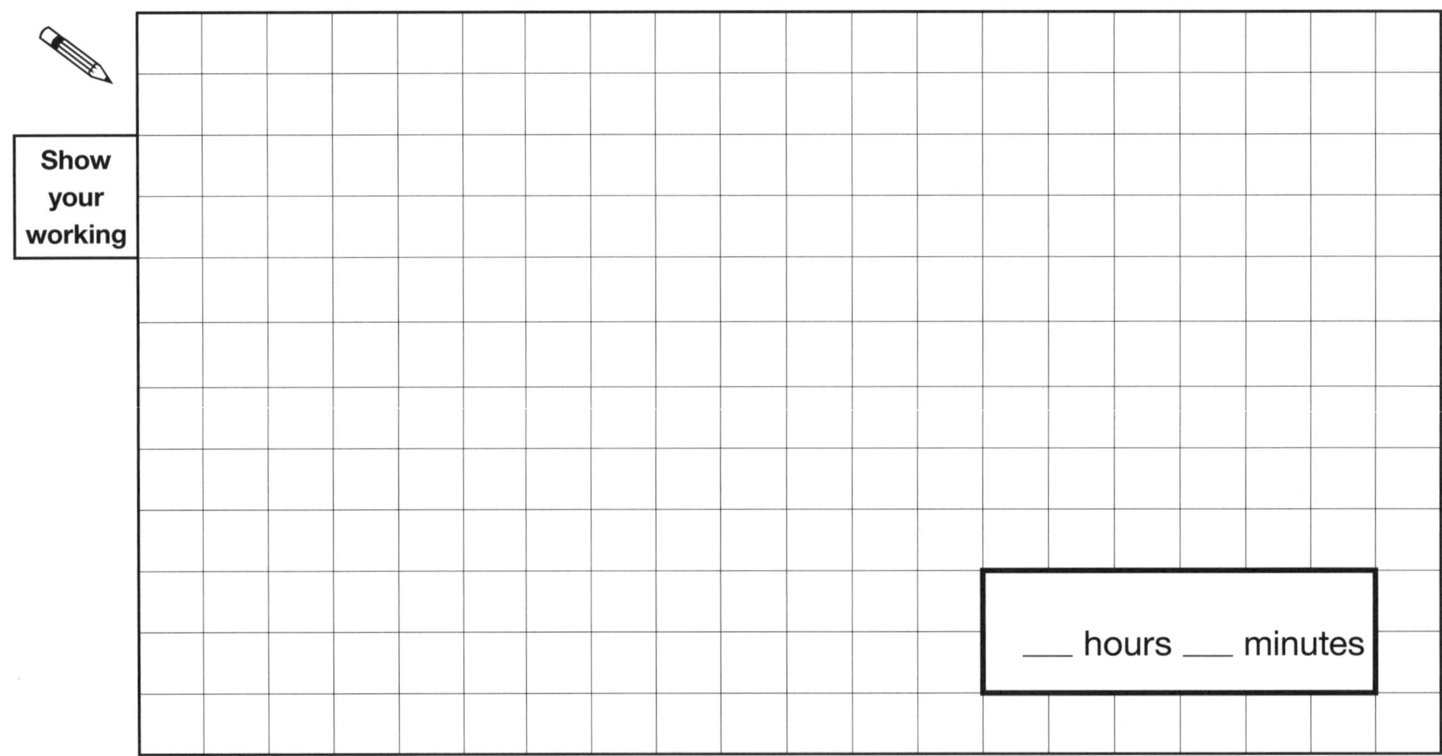

2 marks

5 The perimeter of this rectangle is 56cm.

The length of the rectangle is 17cm.

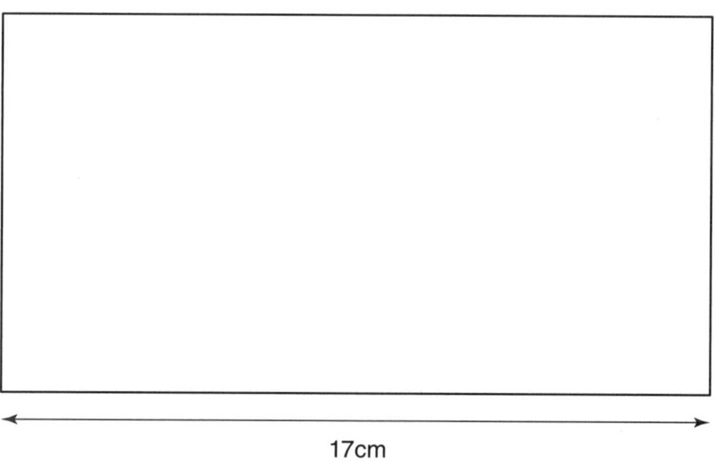

Calculate the width of the rectangle.

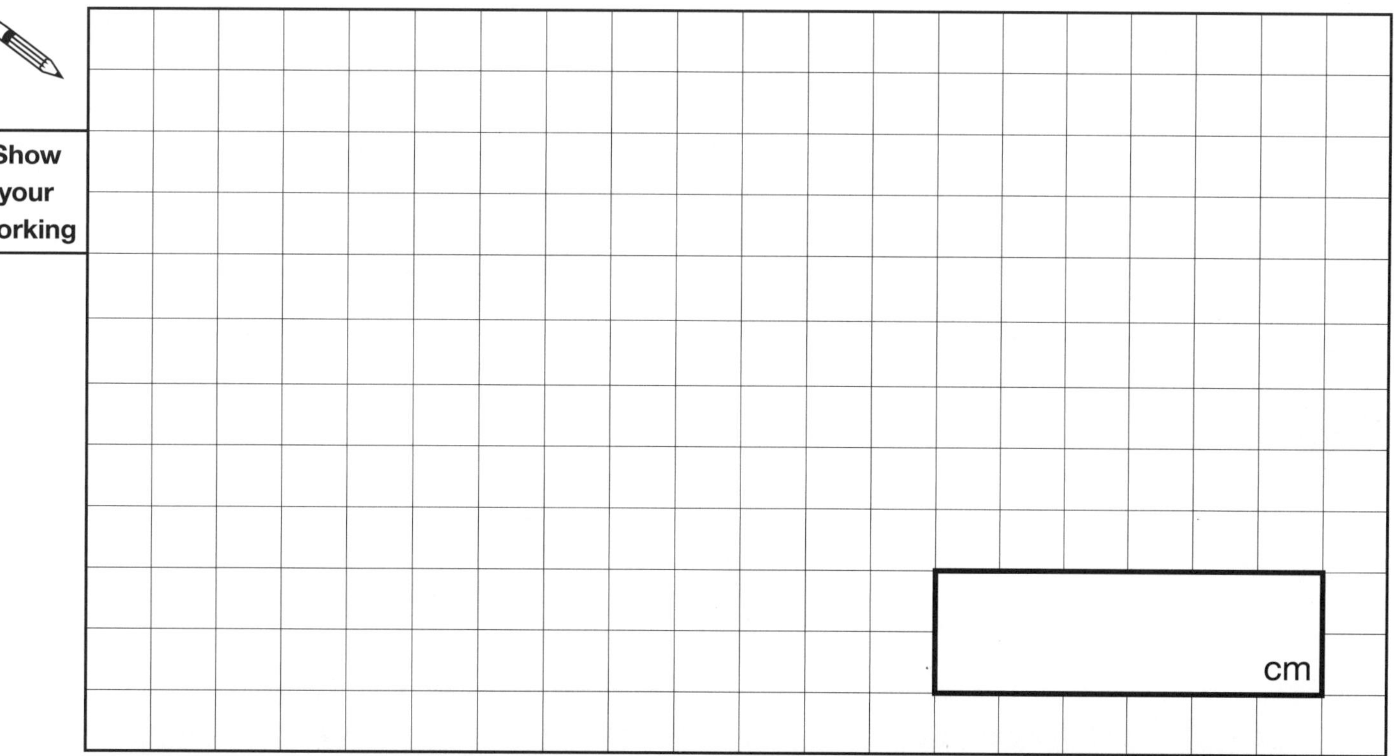

2 marks

6 Simplify $\frac{16}{24}$

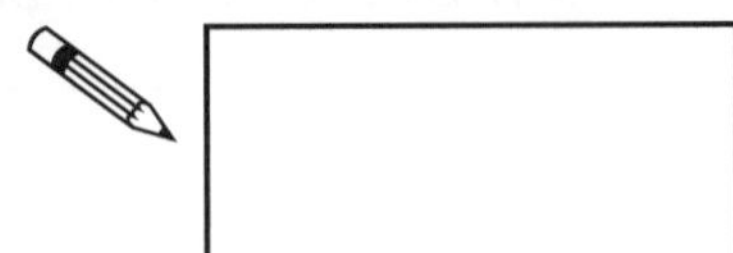

1 mark

7 Four friends share the cost of a meal equally.

The meal costs £63

How much do they each pay?

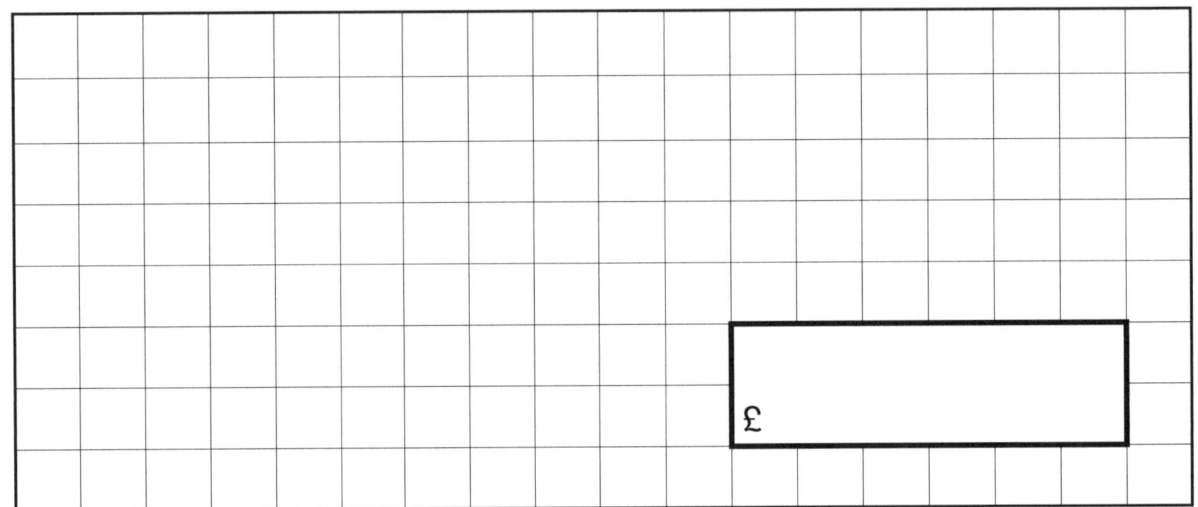

1 mark

8

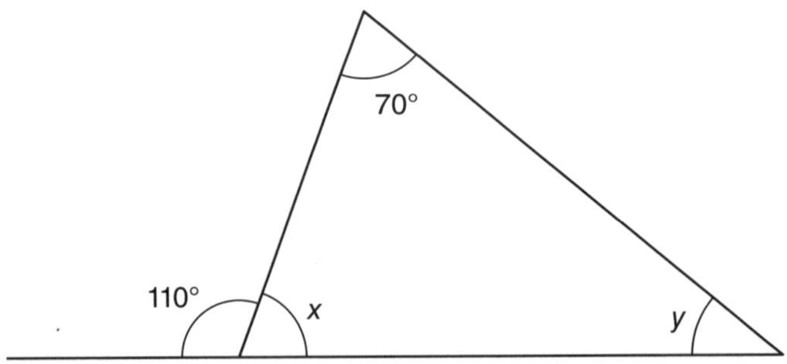

Calculate the missing angles.

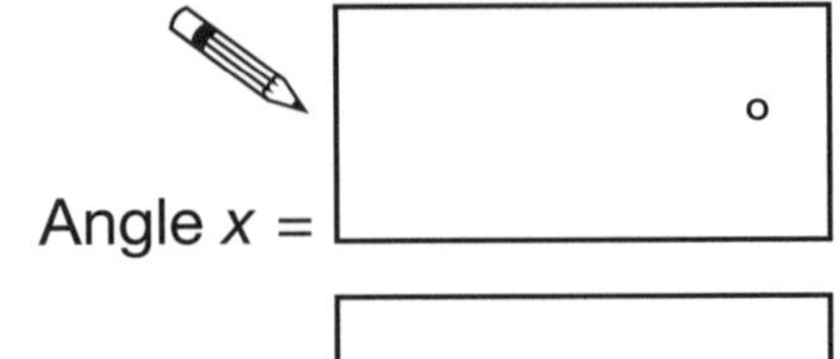

Angle x = °

Angle y = °

2 marks

9 A car hire firm charges its customers, in pounds (£), by the day and for every mile driven.

They use the formula $10d + 0.1m$

where d = the number of days and m = the number of miles

How much would the car hire firm charge for renting a car for 3 days and driving 50 miles?

Show your working

£

2 marks

Total ________ / 15 marks

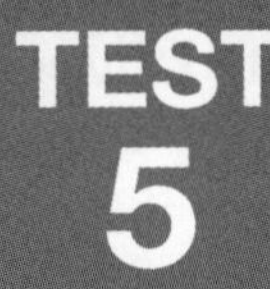

Using Mathematics

- You have 20 minutes to complete this test.
- Calculator not allowed.

1 This bar chart shows the number of minutes Mohammed spends on his mobile phone in one week.

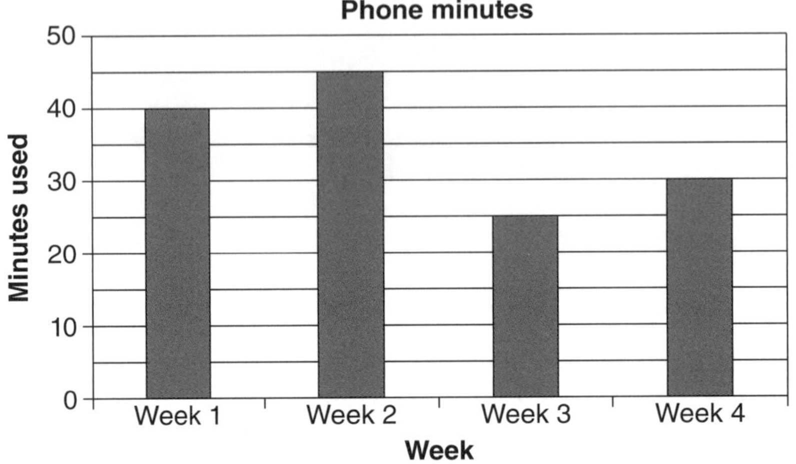

Mohammed's mobile phone contract allows him 200 minutes for four weeks.

How many minutes does Mohammed have left?

Show your working

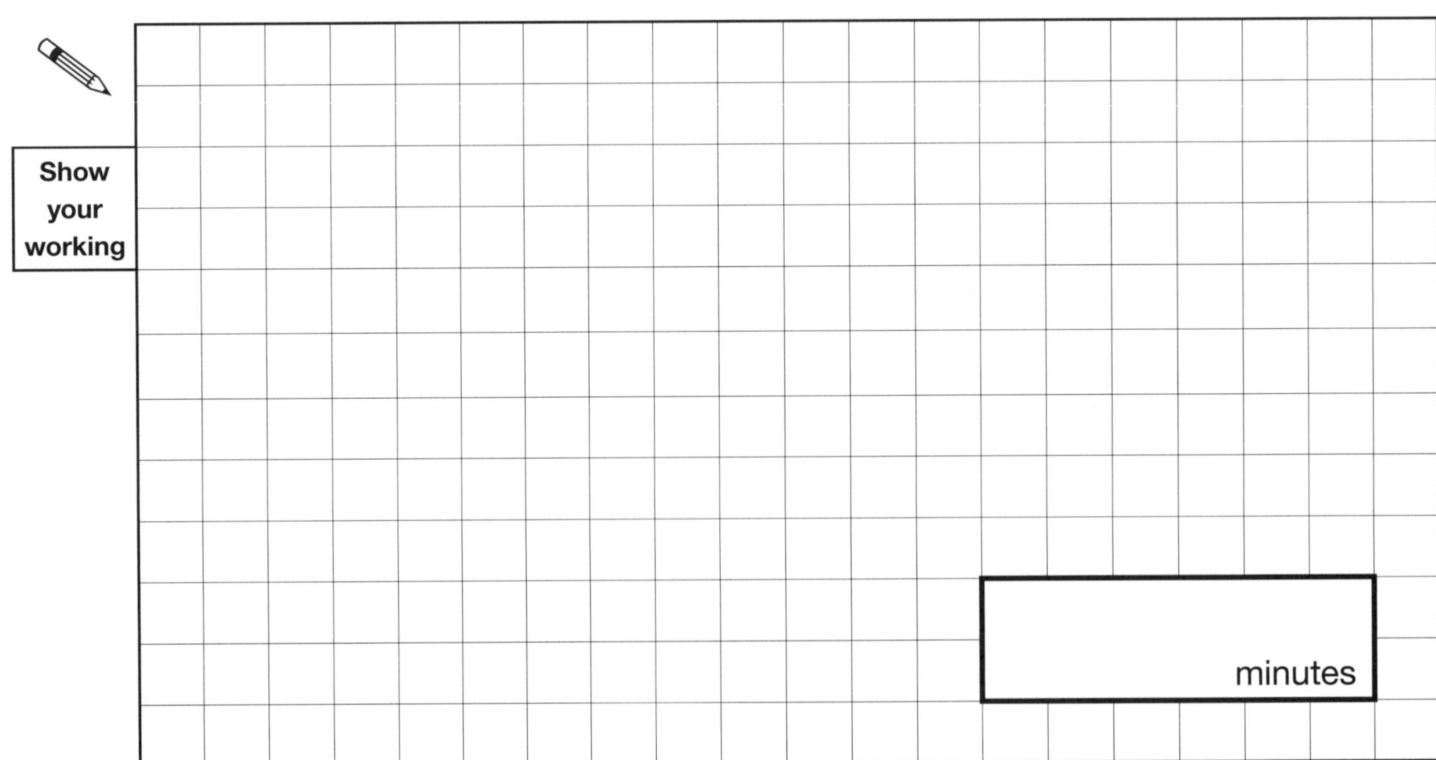

2 marks

2 Sonny uses two £20 notes to buy four books that cost £6.99 each.

How much change does he get?

£

1 mark

3 Tick (✔) the number which has a 3 with the value of thirty thousand.

☐ **3**47 268

☐ 876 0**3**8

☐ 9**3**1 246

☐ 74**3** 971

☐ 275 **3**84

1 mark

4 Dev sat three tests.

His results were:

Maths $\frac{17}{20}$

English $\frac{4}{5}$

Science $\frac{42}{50}$

Dev uses percentages to compare his results.

What was his best result as a percentage?

%

1 mark

5 Once a month, a lorry delivers a load of goods to a shop.

The whole journey is 376 miles.

a. How many miles does the lorry travel over 12 months?

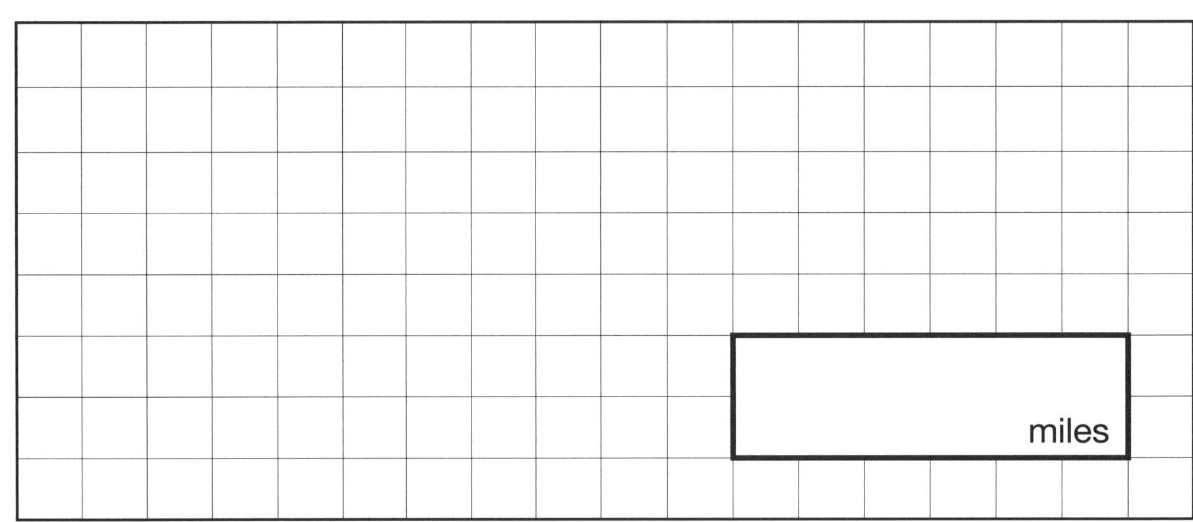

1 mark

b. The lorry uses a gallon of fuel every 8 miles.

How many gallons does the lorry use on **each** journey?

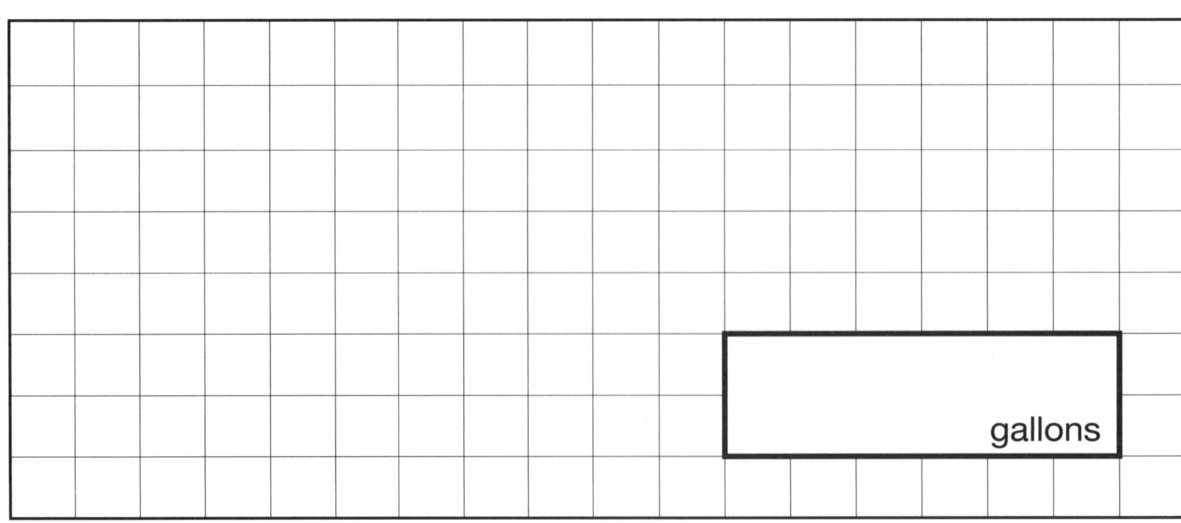

1 mark

6 Max makes some concrete for a path.

For a path 8 metres long Max needs:

- 200kg of cement
- 600kg of sand
- 600kg of stone

What weight of stone will he need for a path 20 metres long?

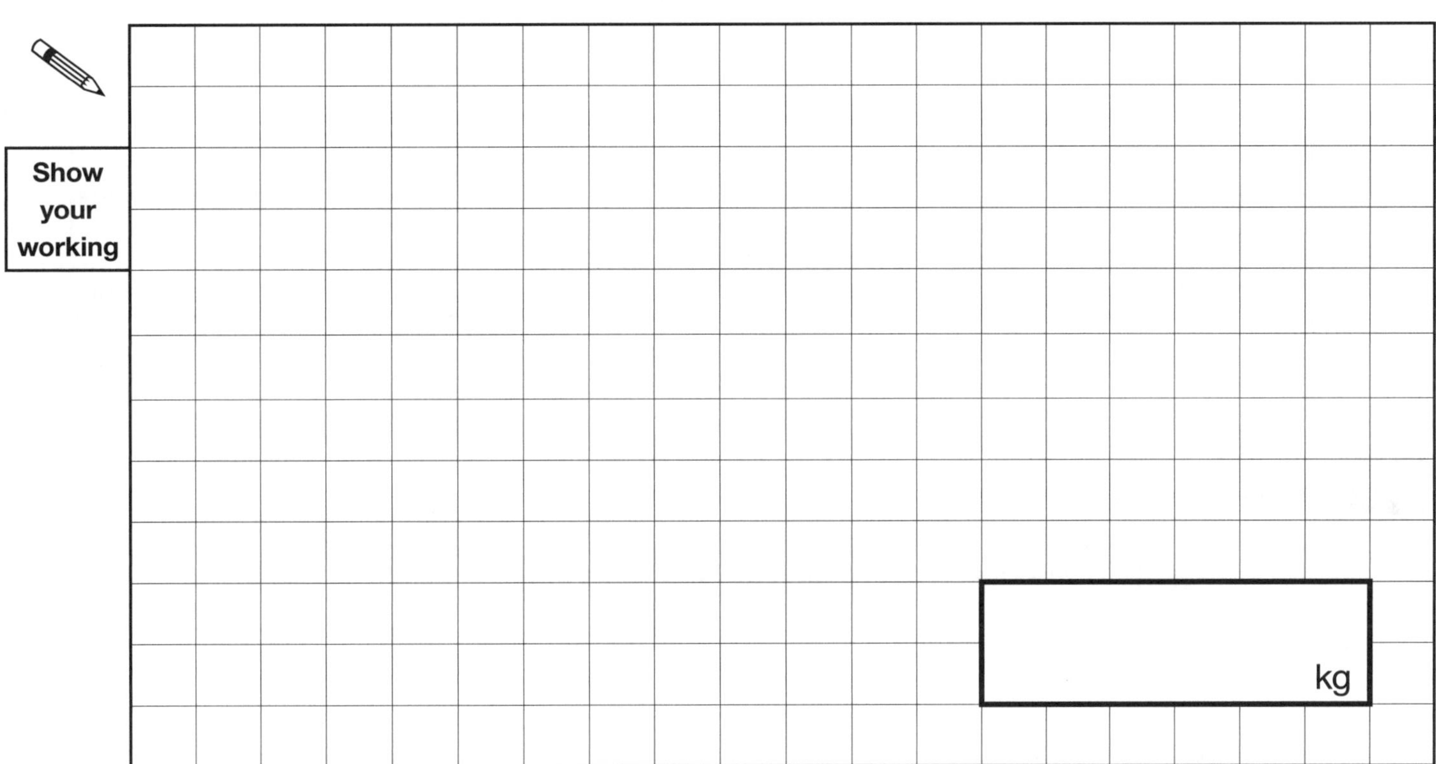

2 marks

7 **Write the missing numbers:**

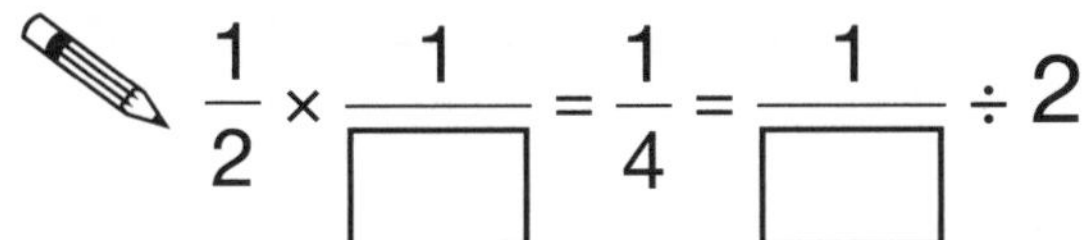

$$\frac{1}{2} \times \frac{1}{\square} = \frac{1}{4} = \frac{1}{\square} \div 2$$

2 marks

8 Manisha makes a drink using orange juice and lemonade.

She uses three times as much lemonade as orange juice.

She makes 600ml of drink.

How much lemonade will she need?

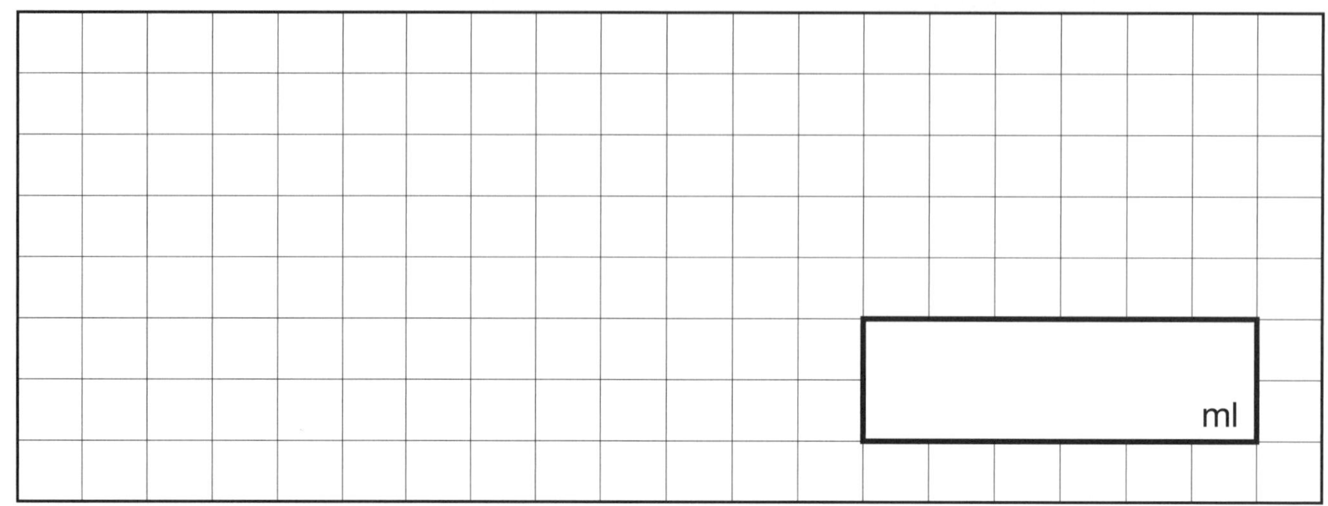

1 mark

9 Approximately, 1 kilometre equals 0.6 miles.

Use this approximation to change 80 kilometres into miles.

1 mark

10 **Complete the labels to name the parts of a circle.**

The centre is marked by a black dot.

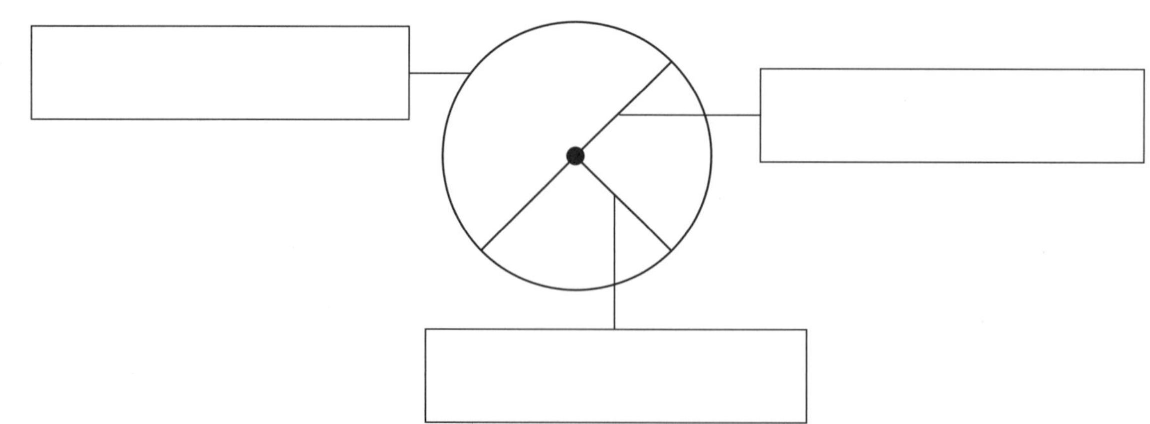

2 marks

Total ________ / 15 marks

Using Mathematics

- You have 20 minutes to complete this test.
- Calculator not allowed.

TEST 6

1 This is part of a calendar.

June						
Sun	**Mon**	**Tue**	**Wed**	**Thu**	**Fri**	**Sat**
			1	2	3	4
5	6	7	8	9	10	11

Emily's birthday was on the last Saturday in May.

What was the date of Emily's birthday?

1 mark

2 Carla looks at a train timetable.

Her train leaves at 16.08

Carla arrives at the station at ten to four.

How long is it before her train leaves?

1 mark

3 The population of a city was 97 358

One hundred years later, the population was 213 064

a. By how many had the population increased?

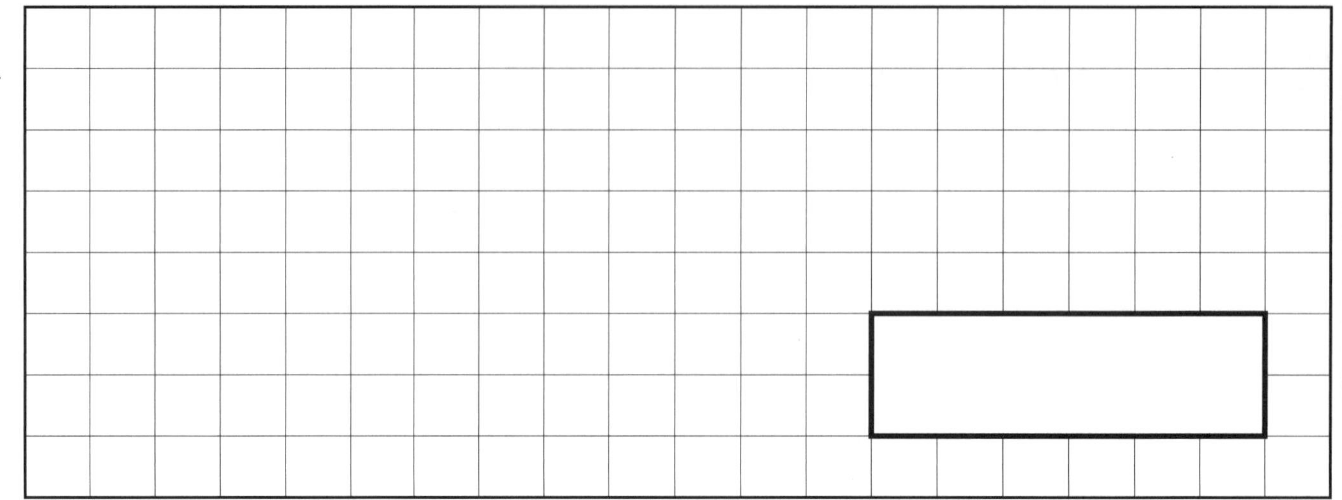

1 mark

b. Another ten years later, the population had increased by another 37 786

What was the population ten years later?

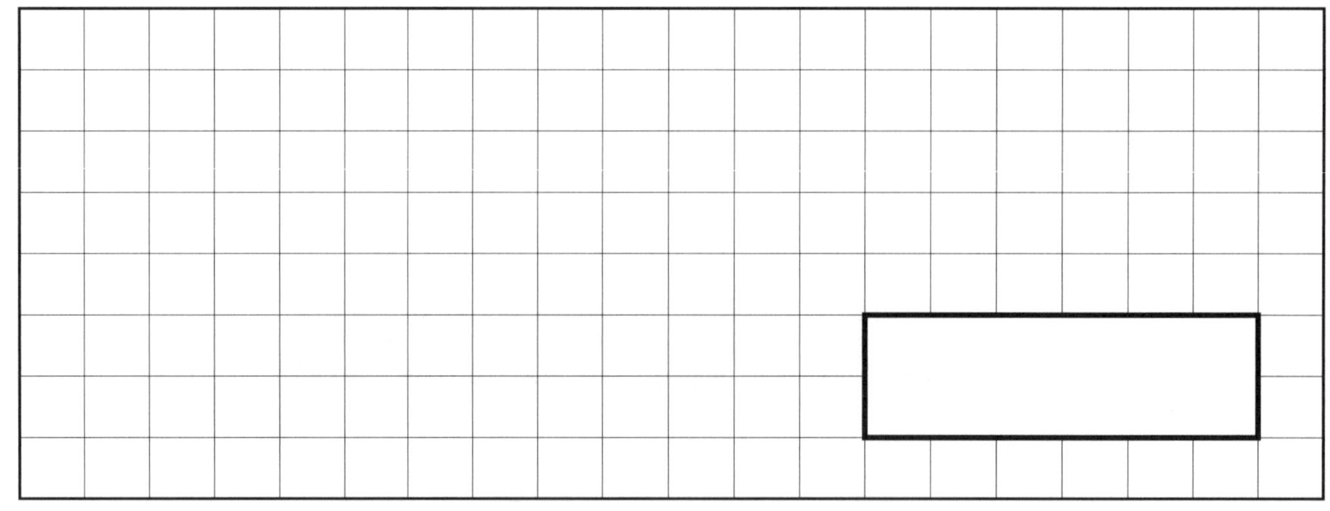

1 mark

4 Put these numbers in order, starting with the largest.

2.903 2.593 2.33 2.9

largest			

1 mark

5 a. Calculate the missing number.

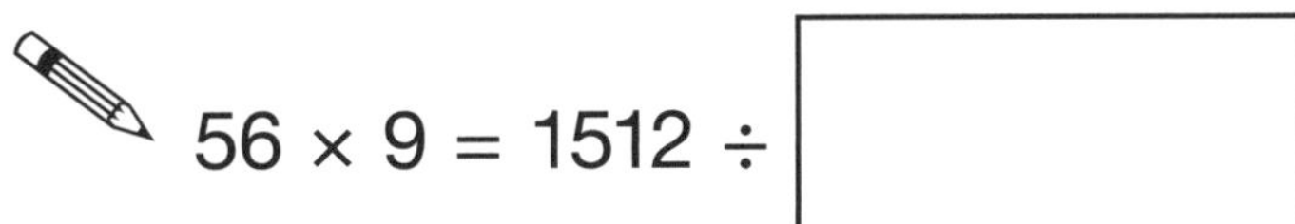

1 mark

b. Calculate the missing number.

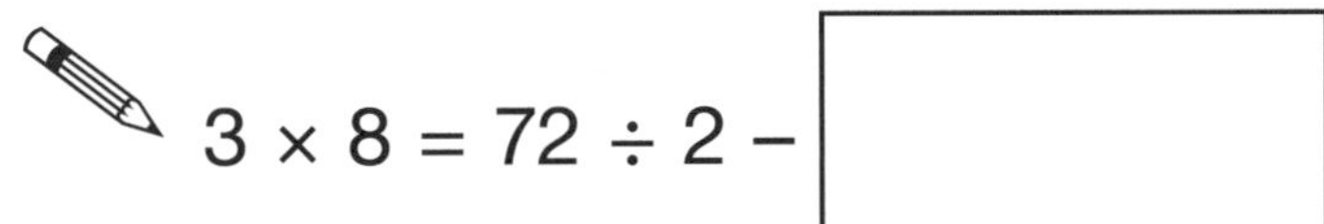

1 mark

6 Calculate the perimeter of this shape.

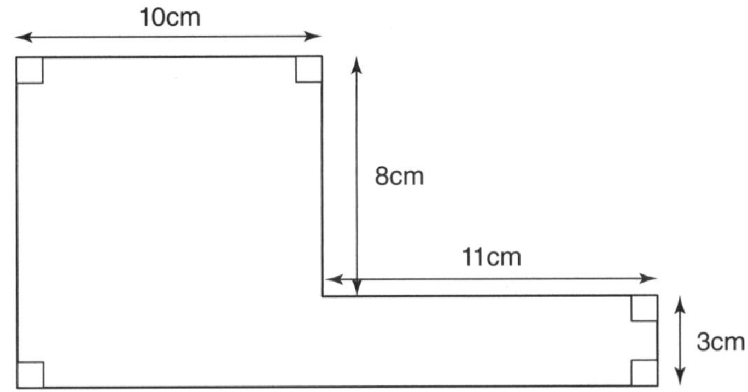

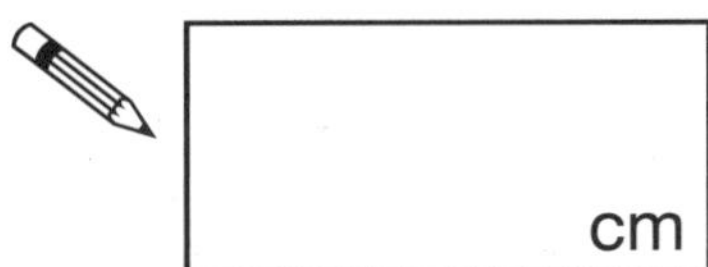

1 mark

7 Calculate angle *a*.

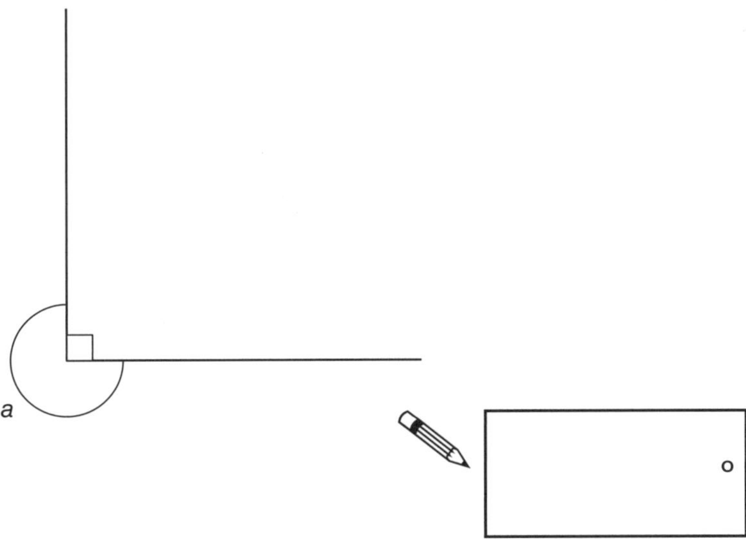

1 mark

8 **a.** Round 319 452 to the nearest hundred.

1 mark

b. Round 319 452 to the nearest hundred thousand.

1 mark

9 Tick (✔) the number that is a common multiple of 12 and 15

3	5	30	60	75
	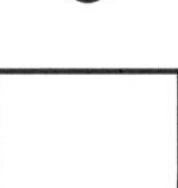	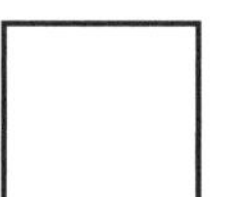		☐

1 mark

10 These shapes stand for numbers.

□ × △ = 36

△ ÷ □ = 4

Which numbers do the shapes stand for?

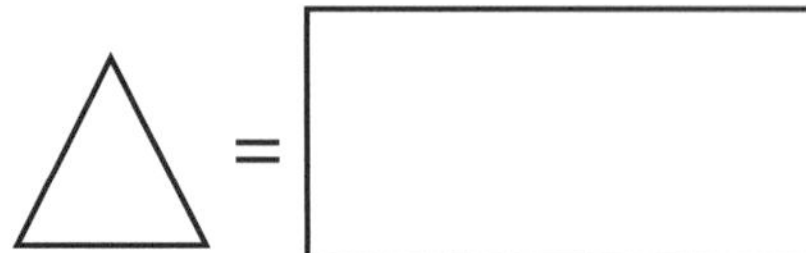

1 mark

11 Two vertices of a square are marked on the grid with the symbol ●

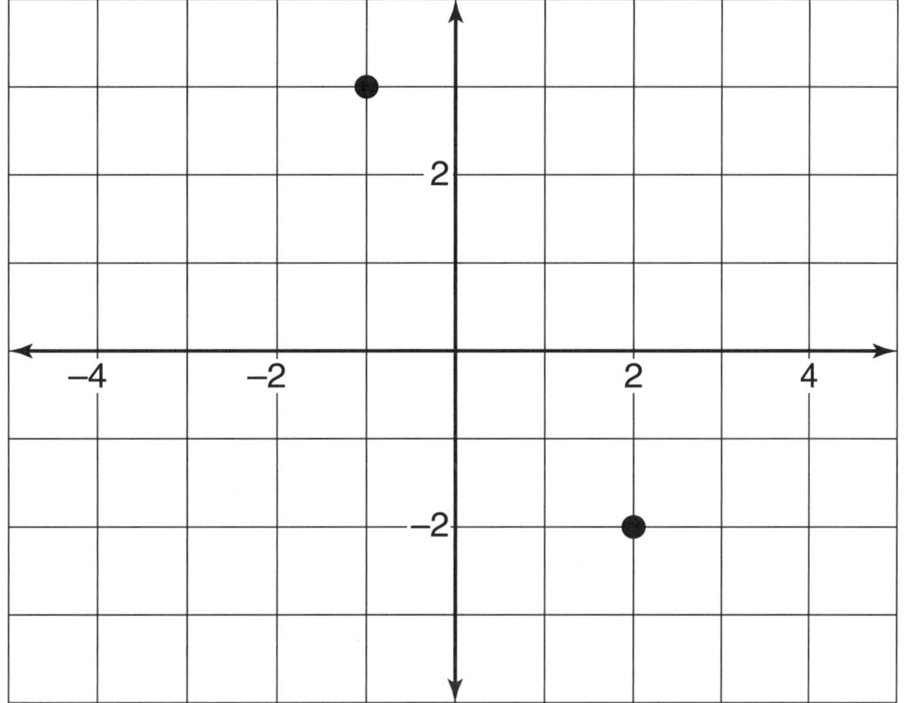

a. Mark a third vertex at (3, 2).

1 mark

b. What are the coordinates of the fourth vertex?

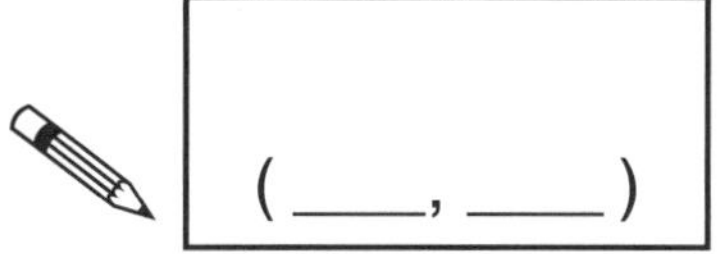

1 mark

12 These fractions are in order of size, smallest first.

Write the missing numbers.

$\frac{7}{12}$ $\frac{\square}{24}$ $\frac{2}{3}$ $\frac{\square}{6}$ $\frac{7}{8}$

2 marks

Total ________ / 17 marks

TEST 7

Using Mathematics

- You have 20 minutes to complete this test.
- Calculator not allowed.

1 Dom sells computer games.

This pictogram shows the number of computer games he sold one week.

Computer Game Sales

◉ stands for four computer games

Sunday	◉ ◉ ◉ ◉
Monday	◉ ◐
Tuesday	◉ ◕
Wednesday	◉
Thursday	◉ ◉ ◉
Friday	◉ ◉ ◉ ◔
Saturday	◉ ◉ ◉ ◉

a. How many computer games did Dom sell on Friday and Saturday?

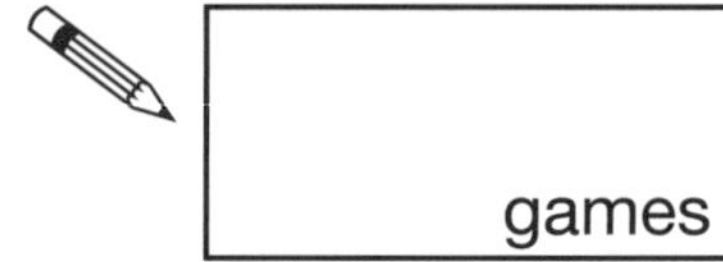

1 mark

b. How many more games did Dom sell on Sunday than on Monday?

1 mark

2 Circle the prime numbers.

 19 29 39 49 59 69

2 marks

3

What time is shown on the clock?

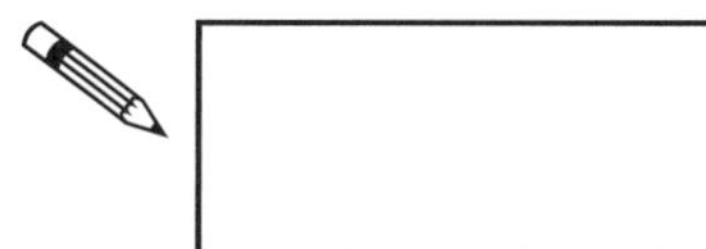

1 mark

4

$\frac{1}{2} + \frac{11}{12} + \frac{3}{4} =$

Give your answer as a mixed number with simplified fraction.

Show your working

3 marks

5

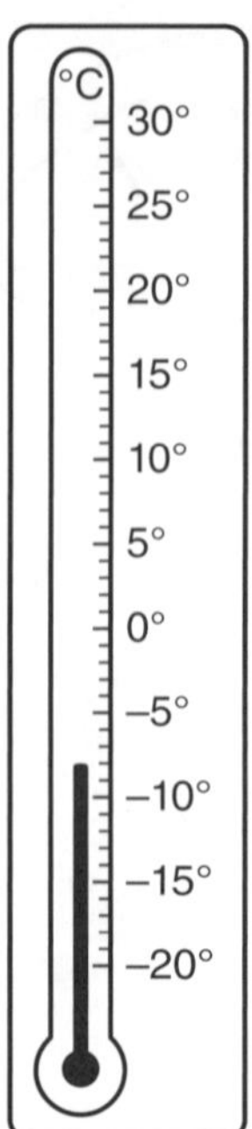

What will the temperature be after the temperature shown on the thermometer has gone up by 3°C?

1 mark

6 Here is part of a number sequence.

The numbers decrease in equal steps.

Fill in the missing numbers.

 58  40  28

2 marks

7 Put these fractions in order, starting with the smallest.

$\frac{3}{4}$ $\frac{5}{6}$ $\frac{1}{2}$ $\frac{7}{12}$ $\frac{2}{3}$

smallest

1 mark

8 Write six million, two hundred thousand, three hundred and fifteen in figures.

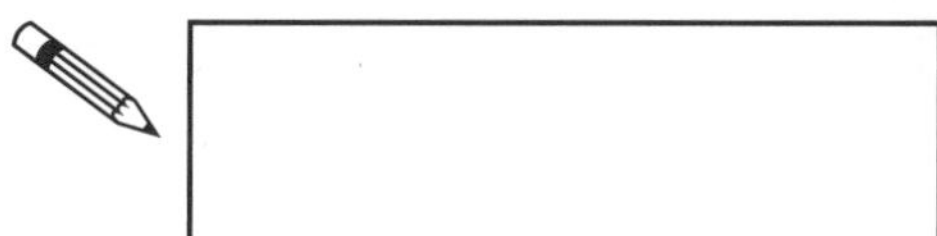

1 mark

9 This shape is a parallelogram.

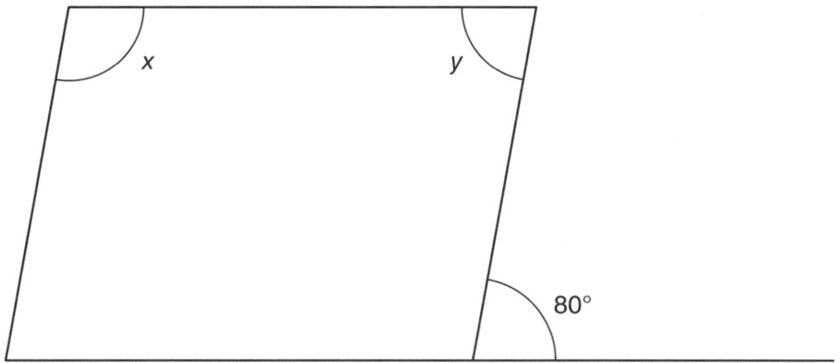

Calculate angle x and angle y.

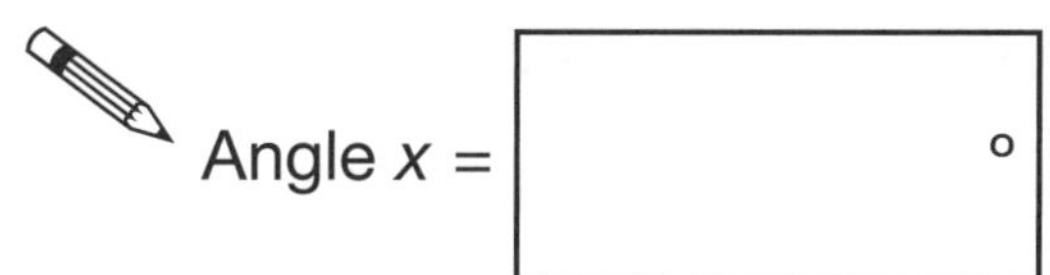

Angle x = °

Angle y = °

2 marks

10 Calculate the missing numbers.

 **a.** 15% of 250 = 

1 mark

b. 15% of ☐ = 90

1 mark

Total ________ / 17 marks

Using Mathematics

- You have 20 minutes to complete this test.
- Calculator not allowed.

1 Ellie has a 2 litre container of juice.

She pours equal amounts of juice into 8 glasses.

How much juice is in each glass?

Give your answer in litres.

1 mark

2 A car firm builds 86 258 cars a year.

35 453 cars are hatchbacks.

27 328 cars are estates.

The rest of the cars are saloons.

How many saloons did the car firm build?

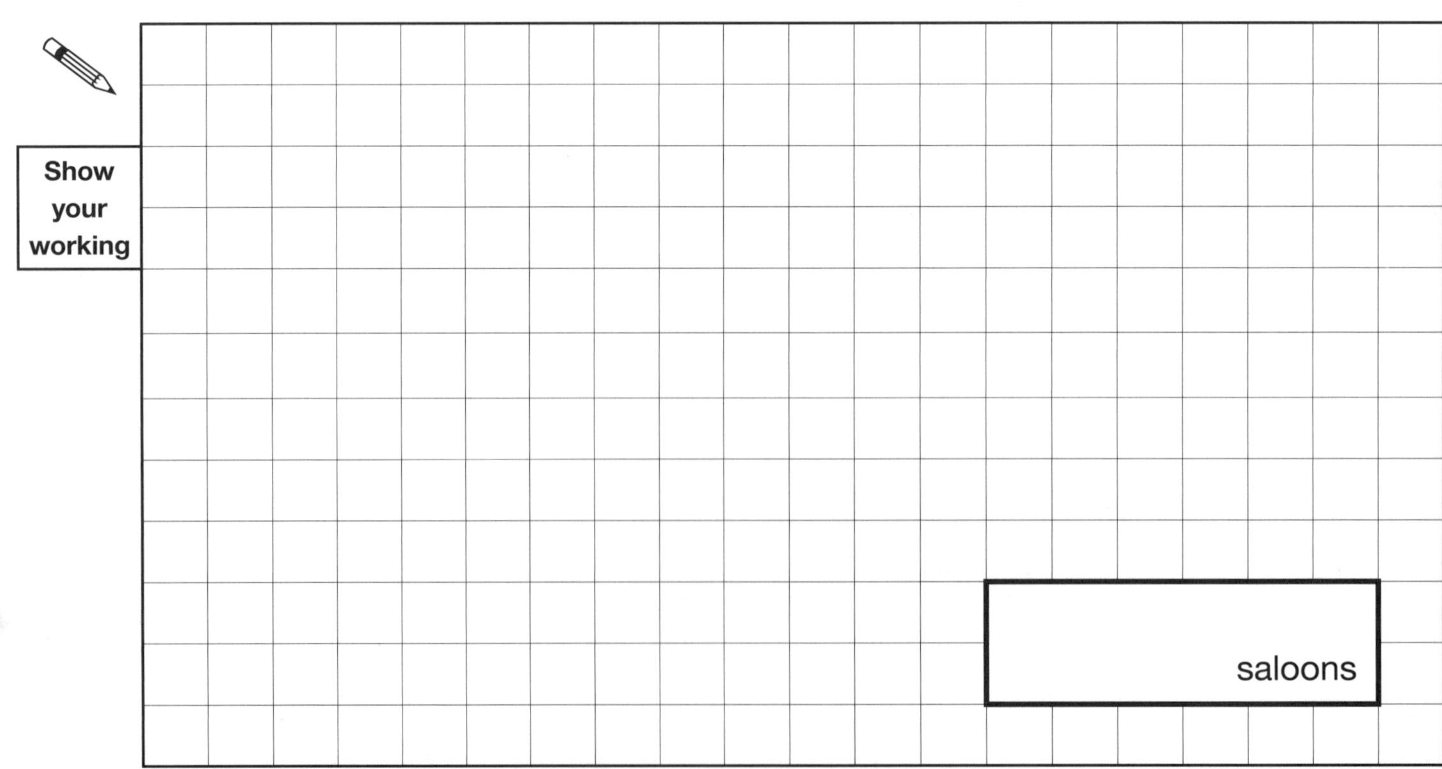

2 marks

3 A factory can print 1632 books every 6 hours.

a. How many books will be printed in 1 hour?

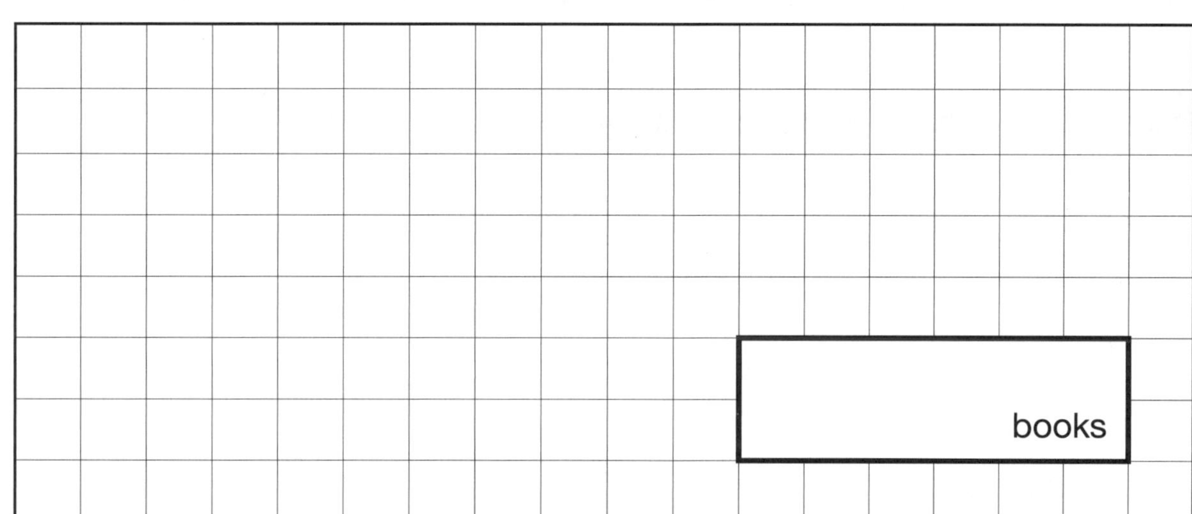

1 mark

b. How many books will be printed in 24 hours?

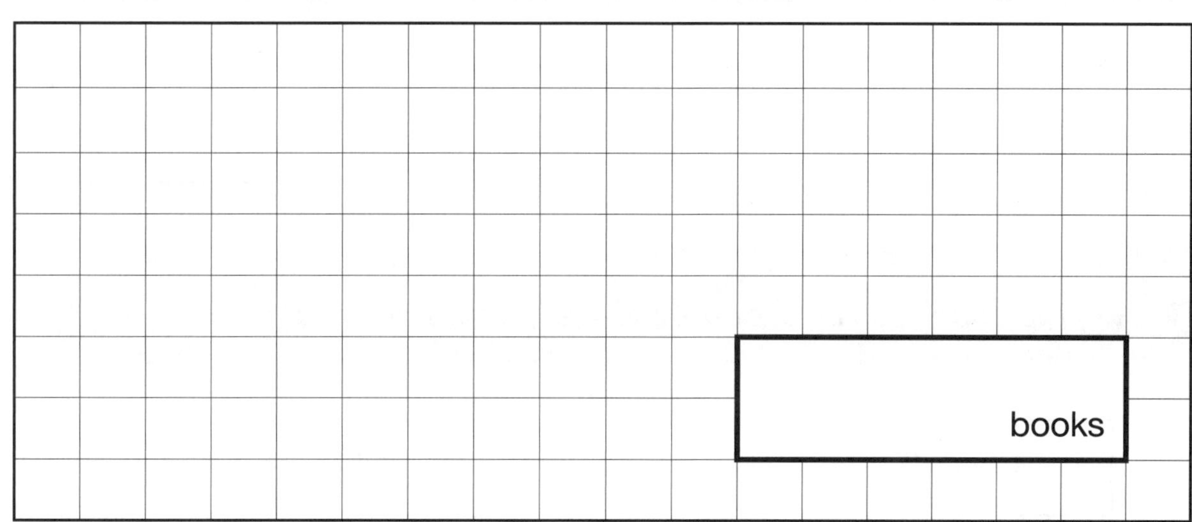

1 mark

4

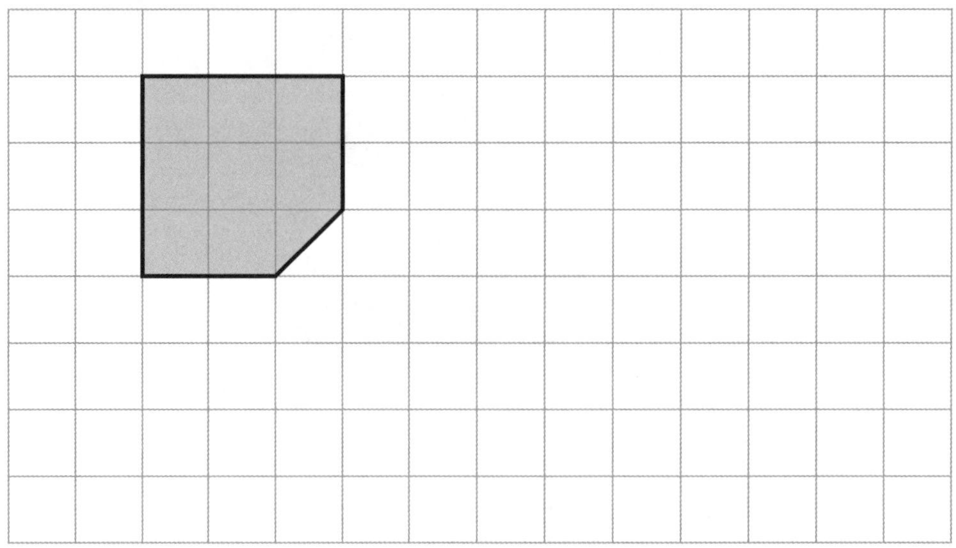

Translate the shape 7 squares right and 3 squares down.

2 marks

5

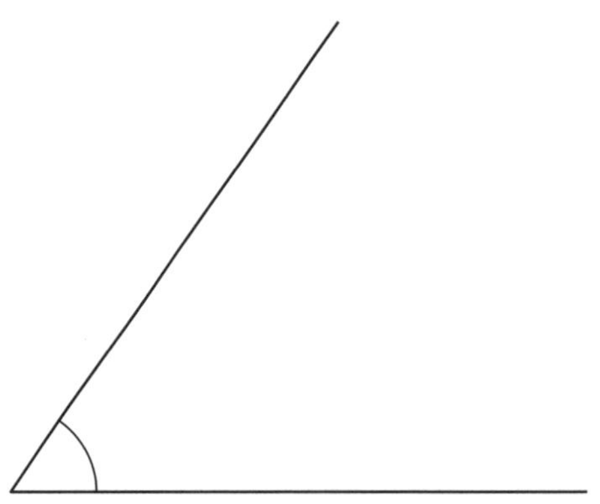

Measure the angle.

1 mark

6

Tick (✔) the correct calculation to change $\frac{2}{5}$ into a decimal.

☐ $5 \div 2 =$

☐ $5 - 2 =$

☐ $2 + 5 =$

☐ $5 \times 2 =$

☐ $2 \div 5 =$

1 mark

7 Ned has a 10kg bag of potatoes.

He uses 2.3kg of potatoes one day.

He uses 1600g of the potatoes on the next day.

What is the weight of potatoes left?

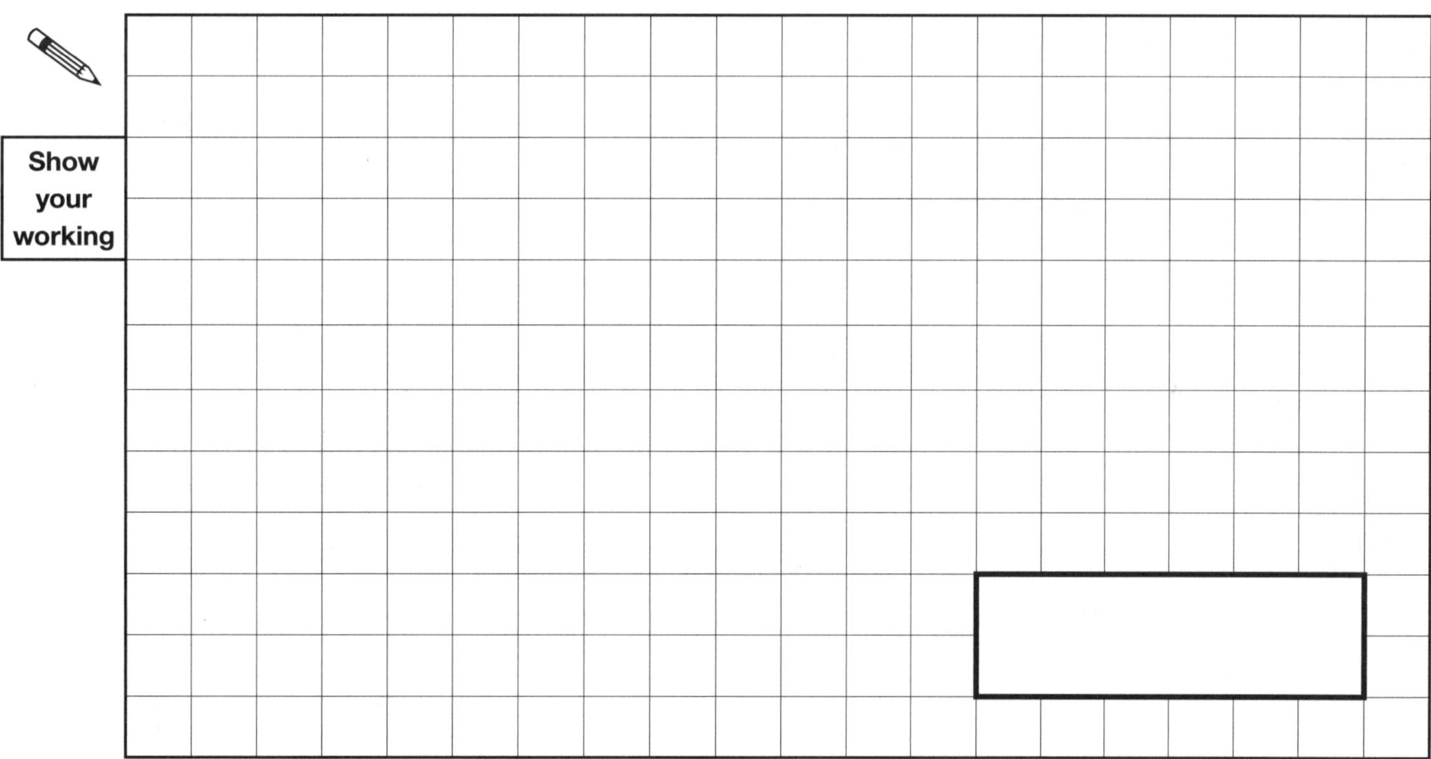

2 marks

8 Write the missing number.

$\frac{1}{3} \div \square = \frac{1}{12}$

1 mark

9 Name the 3-D shape this net will make.

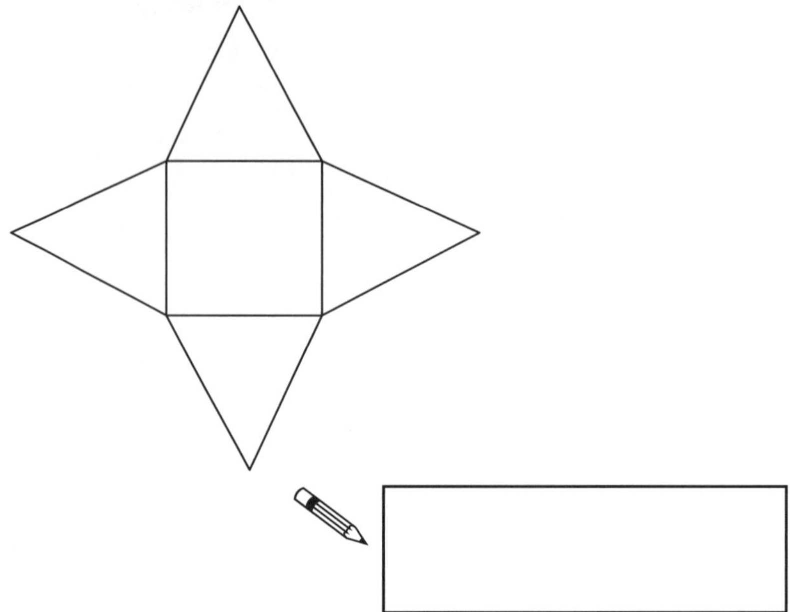

1 mark

10 A factory has 2 machines that bottle drinks.

One machine can fill 2365 bottles in one hour.

A newer machine can fill 3850 bottles in one hour.

The bottles are put into boxes holding 24 bottles.

How many boxes are completely filled in one hour?

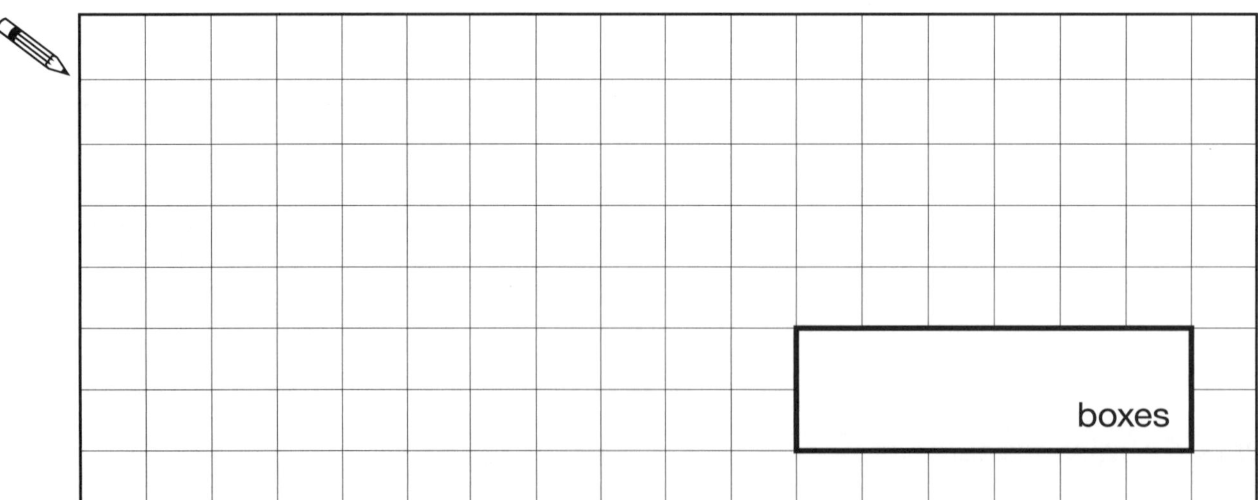

2 marks

Total ________ / 15 marks

Using Mathematics

TEST 9

- You have 20 minutes to complete this test.
- Calculator not allowed.

1 48 boxes can fit into 5 vans.

How many boxes can fit into 20 vans?

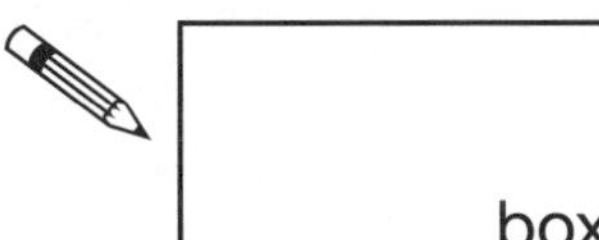

1 mark

2 Complete the missing numbers.

$78 \div 6 = (\square \div 6) + (18 \div \square) = 13$

1 mark

3 Draw lines to join a decimal with its equivalent fraction.

1.3	$\frac{13}{100}$
0.103	$\frac{13}{10}$
0.13	$\frac{103}{1000}$
0.013	$\frac{103}{100}$
1.03	$\frac{13}{1000}$

2 marks

4 Outside temperatures were recorded at midday for a week.

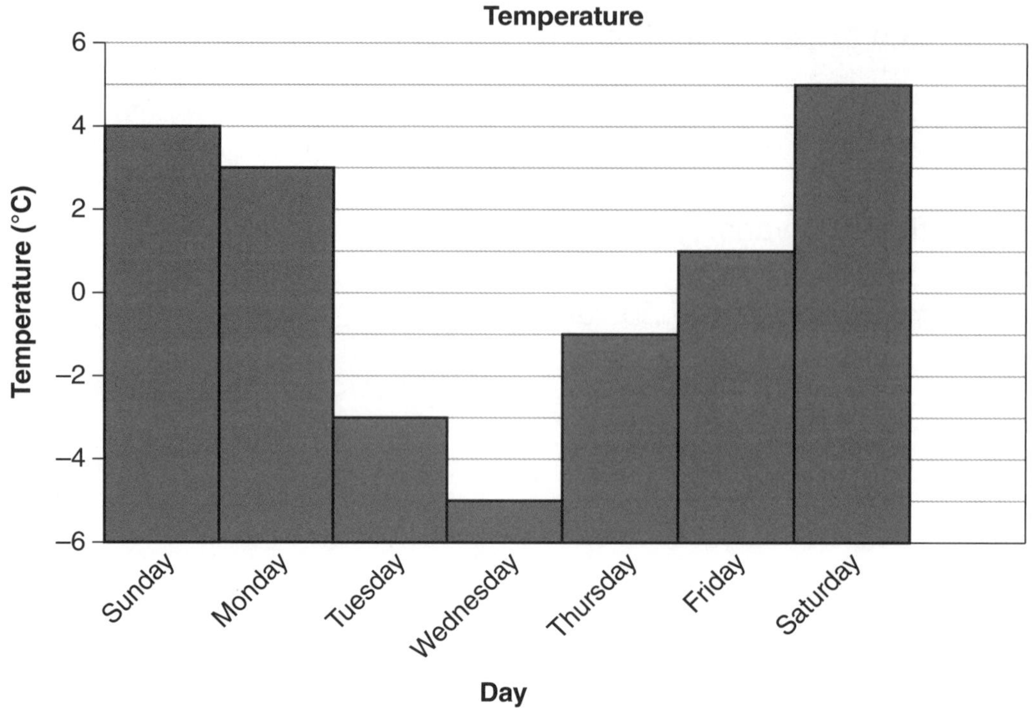

a. What was the temperature at midday on Monday?

1 mark

b. What was the difference between the temperatures recorded on Sunday and Wednesday?

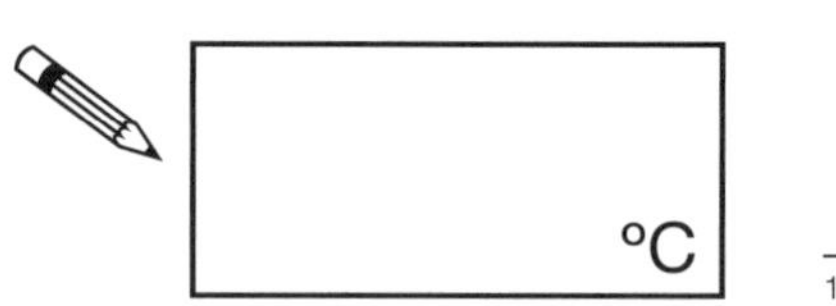

1 mark

c. For how many days was the temperature below 0°C?

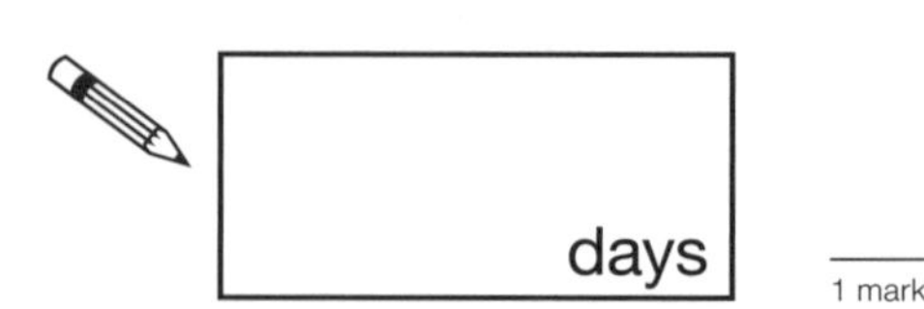

1 mark

5

7 569 135

Milly makes some changes to this number.

She writes:

- 4 as the tens of thousands digit
- 2 as the thousands digit
- 8 as the millions digit

She leaves the other numbers as they are.

Write Milly's new number.

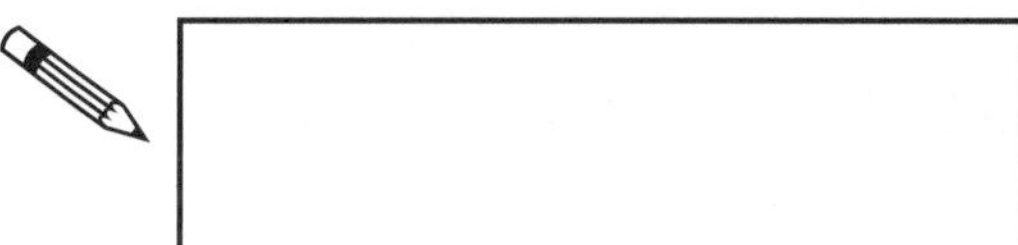

1 mark

6

What is the difference between the area of the square and the area of the oblong?

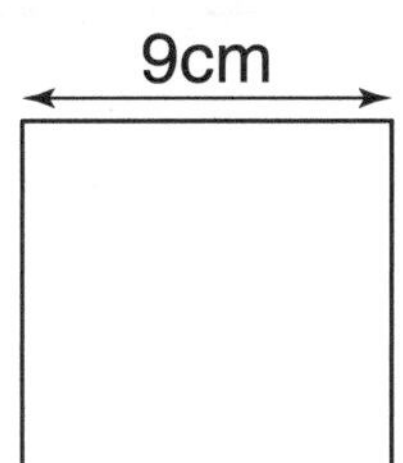

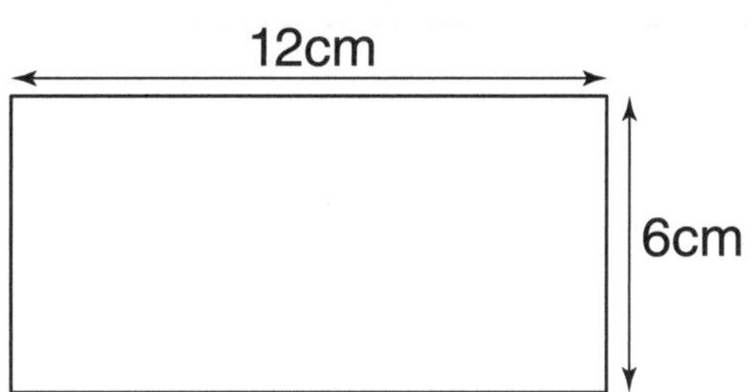

2 marks

7

Calculate the missing number.

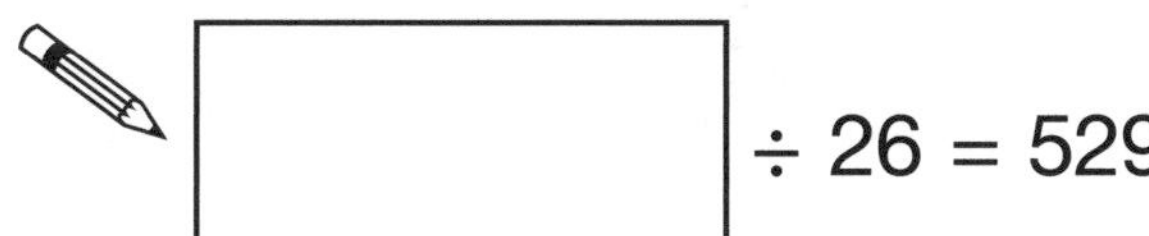

1 mark

8 Calculate

a. 20% of 600 = ☐

1 mark

b. 90% of ☐ = 270

1 mark

c. ☐ % of 200 = 50

1 mark

9 Jo sits 6 maths tests.

Her marks were:

14 16 14 17 19 16

What was her mean mark?

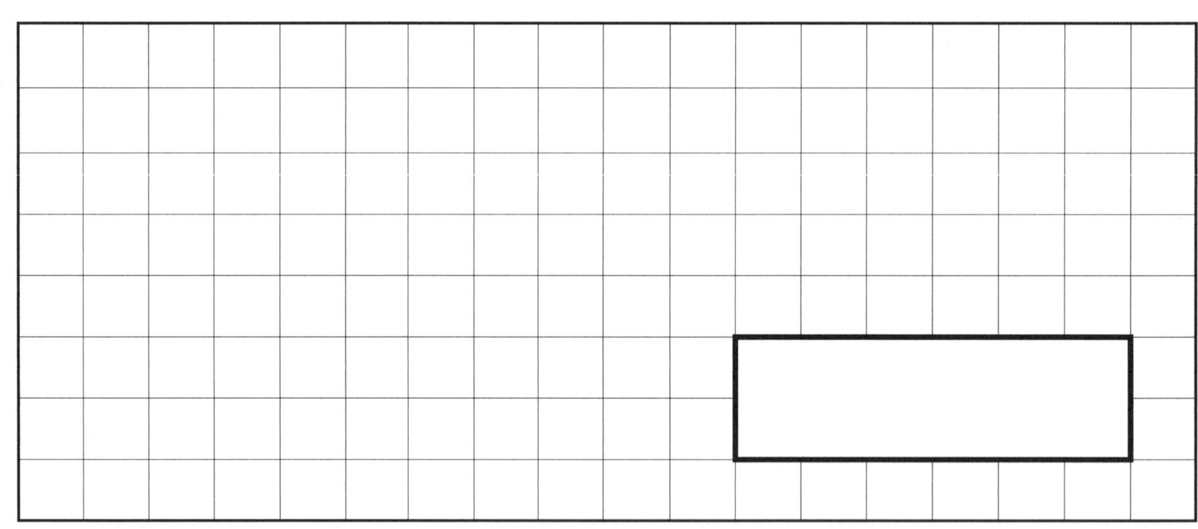

1 mark

Total ________ / 15 marks

Using Mathematics

TEST 10

- You have 20 minutes to complete this test.
- Calculator not allowed.

1 Here are some number cards.

0	5	5	6	9

Rearrange the number cards to make the closest number to 60 000

1 mark

2 Write the missing number.

$5.156 + \square = 12.48$

1 mark

3 Write the missing numbers.

$70\% = \frac{7}{\square}$

$\square\% = \frac{3}{4}$

$80\% = \frac{\square}{5}$

2 marks

4 The dotted line is the mirror line.

Reflect the shape in the mirror line.

Use a ruler.

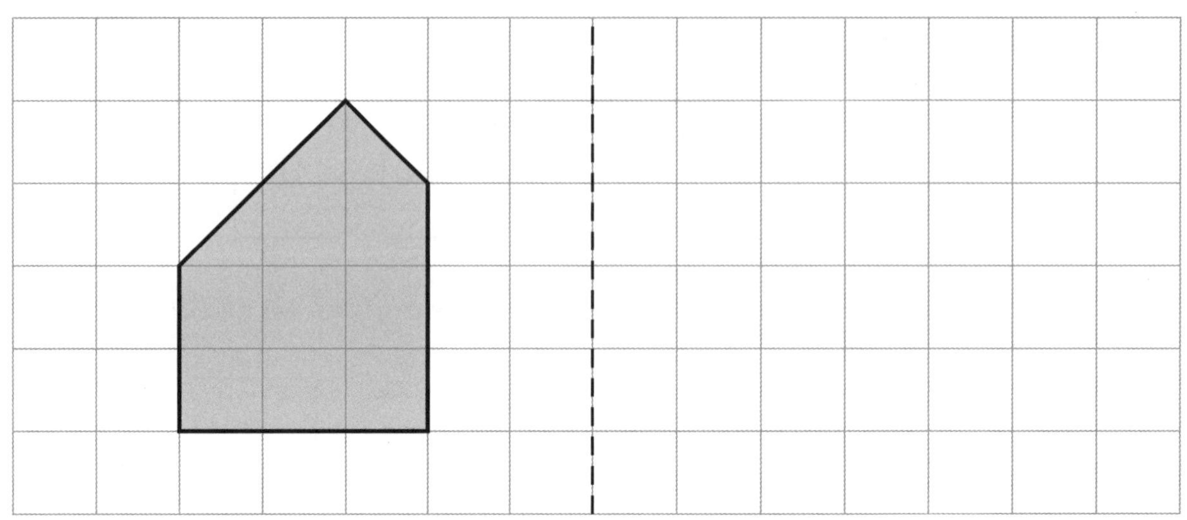

2 marks

5 **What could the missing numbers be?**

$$\frac{1}{\square} \times \frac{1}{\square} = \frac{1}{12}$$

1 mark

6 **Find the missing numbers.**

a. 3.2 ÷ 100 = ☐ ÷ 1000

1 mark

b. 4.7 × 10 = ☐ × 1000

1 mark

7 Debbie runs a cake shop.

She works out the cost of a cake, in pounds (£), by doubling the cost of the ingredients, i, and adding 4.

Tick (✔) the formula Debbie uses.

☐ $2 \times 4 + i$

☐ $2(4 + i)$

☐ $2i + 4$

☐ $4 + i + 2$

☐ $4i + 2$

1 mark

8 A rectangle is 12cm long and 5cm wide.

The rectangle is enlarged by a scale factor of 4

What are the length and width of the enlarged rectangle?

length = ________ cm, width = ________ cm

1 mark

9

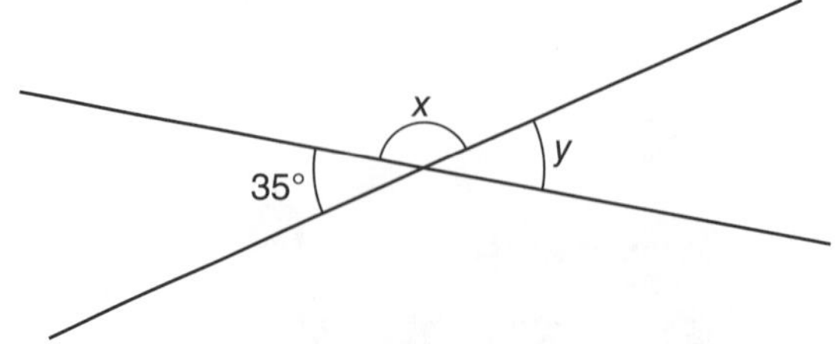

Calculate angle x and angle y.

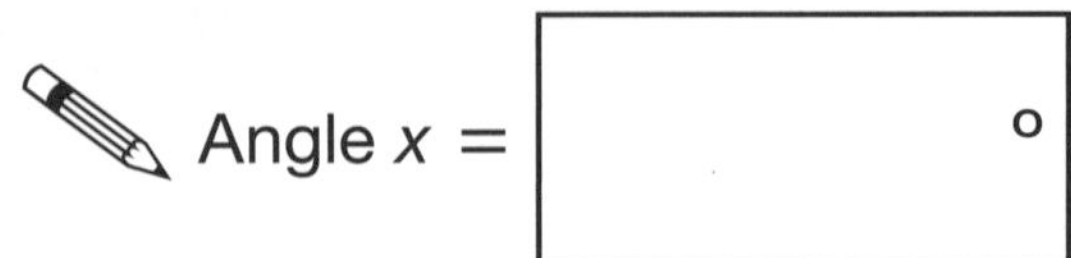

Angle y = ____ °

2 marks

10

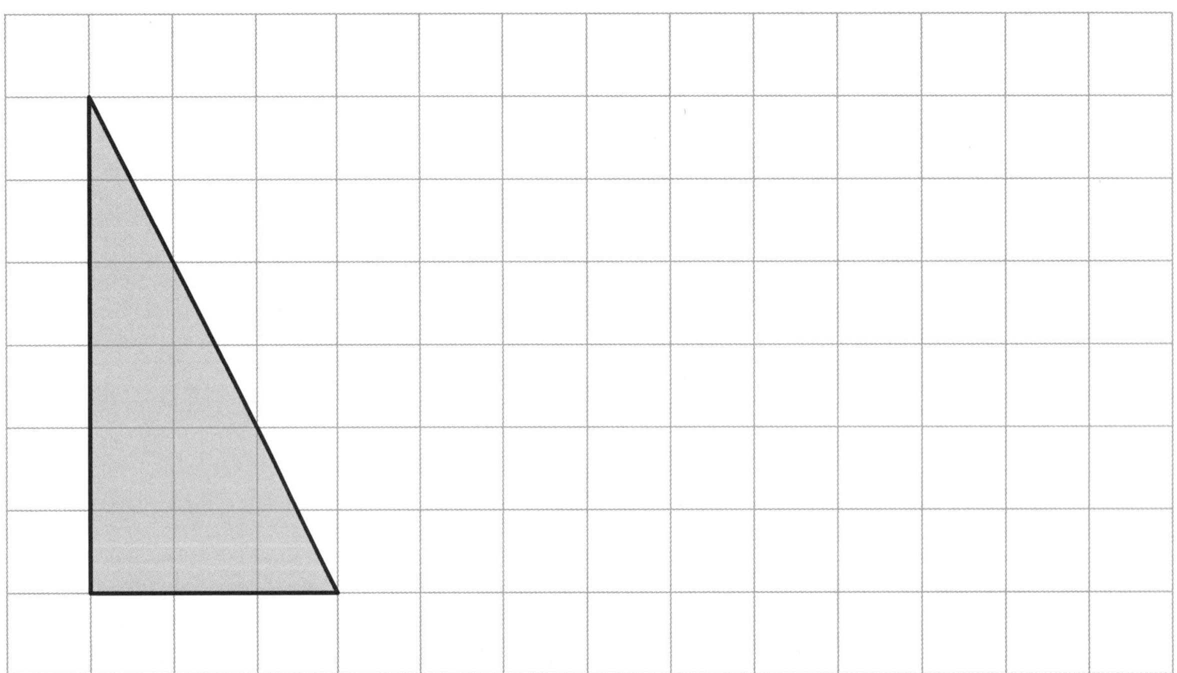

a. Calculate the area of the triangle.

1 mark

b. Draw a **square** on the grid with the same area as the triangle.

Use a ruler.

1 mark

Total ________ / 15 marks

Using Mathematics

TEST 11

- You have 20 minutes to complete this test.
- Calculator not allowed.

1 Ola adds two fractions.

The answer is $1\frac{1}{4}$

Tick (✔) the two fractions Ola adds.

☐ $\frac{1}{4} + \frac{1}{4}$

☐ $\frac{4}{5} + \frac{2}{5}$

☐ $\frac{9}{12} + \frac{7}{12}$

☐ $\frac{3}{8} + \frac{7}{8}$

☐ $\frac{7}{20} + \frac{17}{20}$

1 mark

2 Circle the number with a 6 as hundreds of thousands **and** a 4 as thousands.

543 619   6 456 752   2 634 891   645 894

1 mark

3 **a.** 7.4 has been rounded to one decimal place.

Tick (✔) the number that was rounded.

7.32 7.35 7.45 7.49

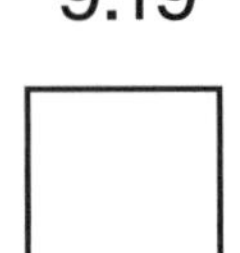

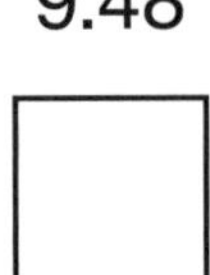

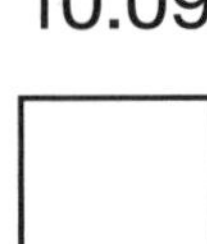

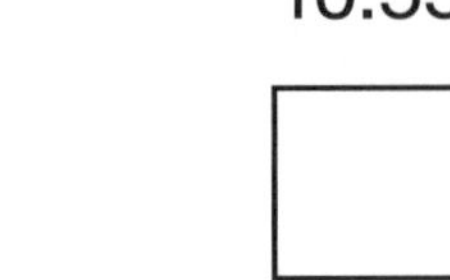

1 mark

b. 10 has been rounded to the nearest whole number.

Tick (✔) the number that was rounded.

9.19 9.48 10.09 10.55

1 mark

4 **Calculate the missing numbers.**

	☐	5	☐	8
−	3	☐	9	☐
	3	4	2	4

1 mark

5 This is a menu for a takeaway.

Meal	Price
Pizza – 12 inch	£9.50
Pizza – 10 inch	£7.00
Fish	£3.20
Large chips	£1.50
Small chips	90p

Judy uses a £20 note to buy:

- a 10 inch pizza
- 2 fish
- 1 large chips
- 1 small chips

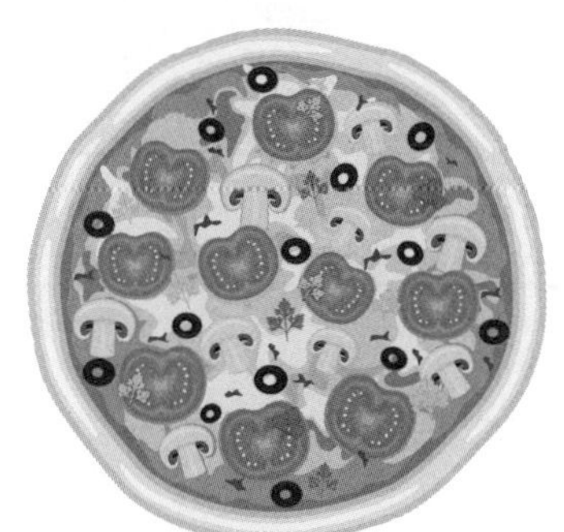

How much change will Judy get?

Show your working

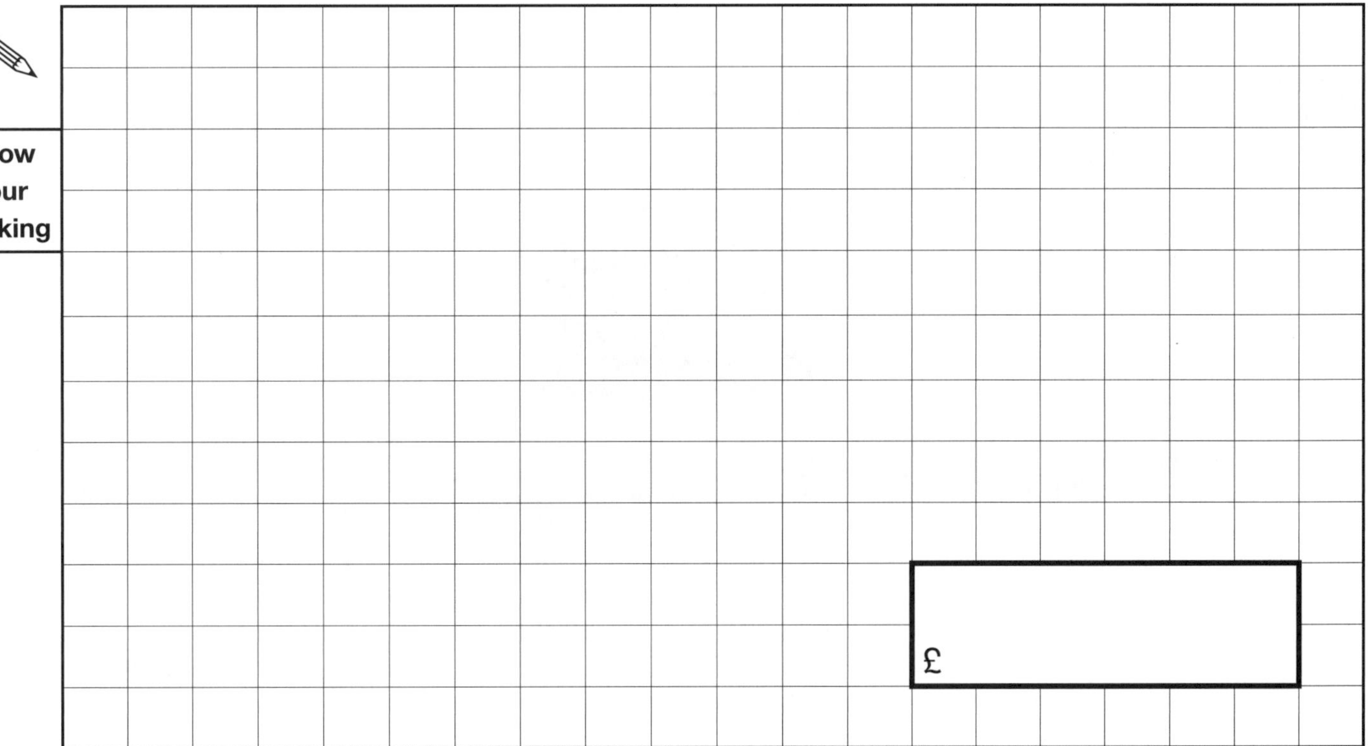

2 marks

6 The two dotted lines are both mirror lines.

A shape has been reflected in both lines.

Draw the two missing shapes from the grid.

Use a ruler.

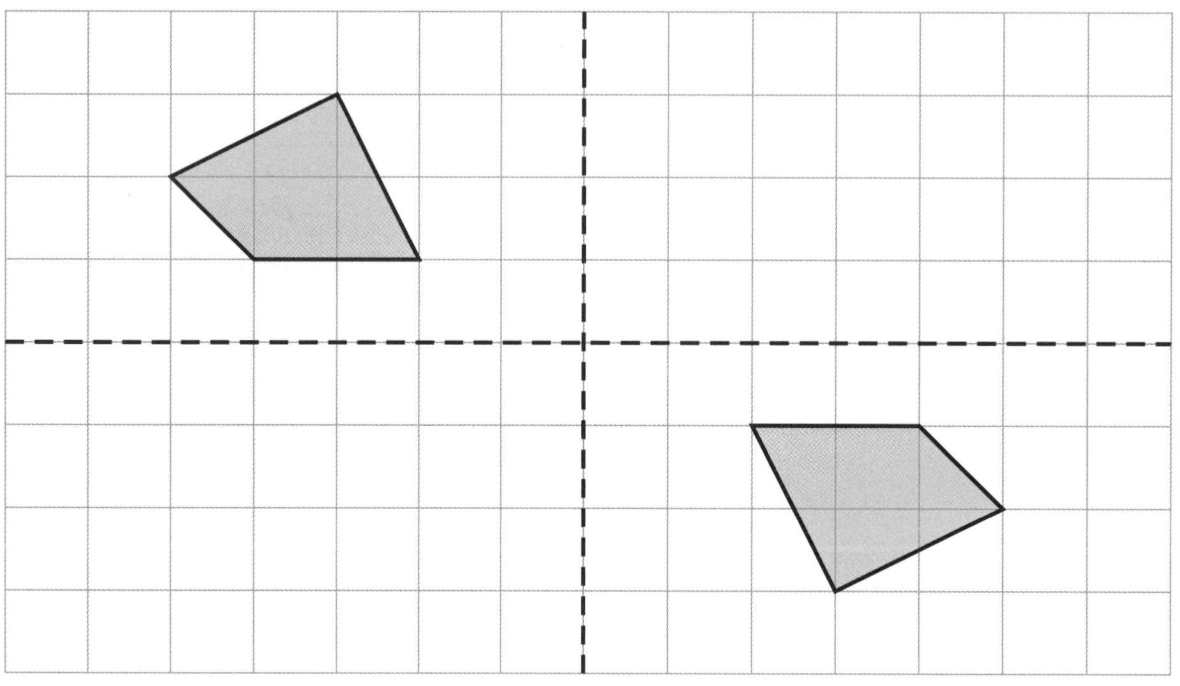

2 marks

7 Fay can buy a 2kg bag of 14 apples for £3

a. **How many apples can Fay buy for £1.50?**

1 mark

b. **How many apples would Fay have if she bought 5kg?**

1 mark

8 Here are two fair triangular spinners.

Each spinner is spun once and the numbers added to give a total.

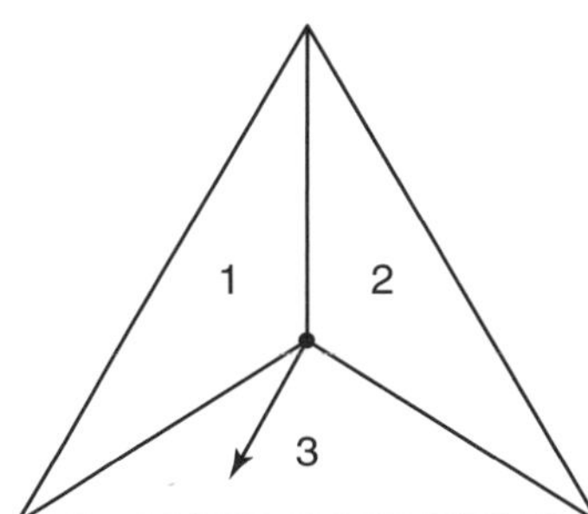

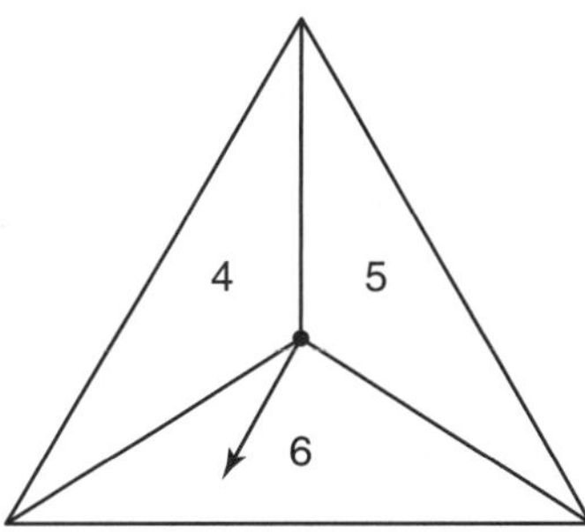

How many different totals can be made?

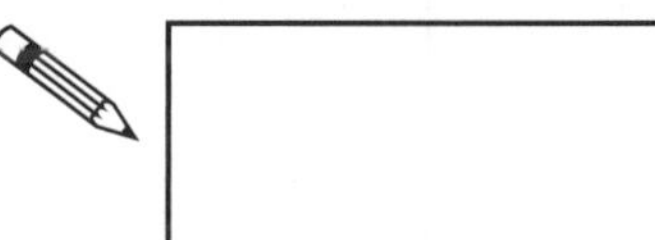

1 mark

9 Find the missing numbers.

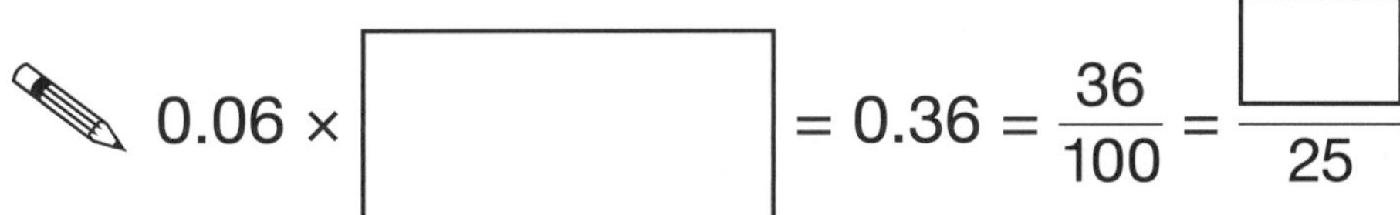

$0.06 \times \square = 0.36 = \frac{36}{100} = \frac{\square}{25}$

2 marks

10 The triangle is an **equilateral** triangle.

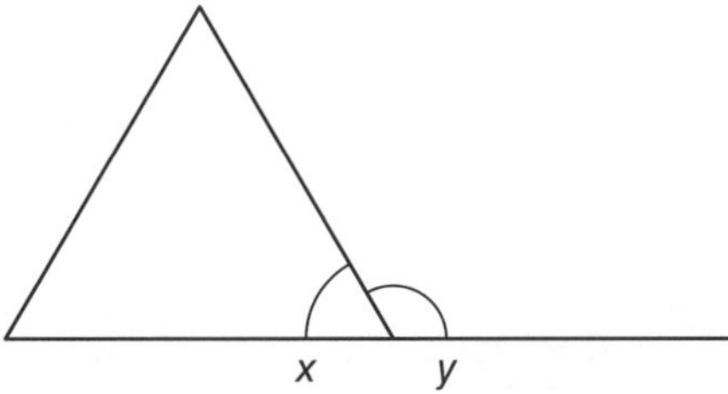

Calculate angle x and angle y.

Angle x = ☐ °

Angle y = ☐ °

2 marks

Total ________ / 16 marks

TEST 12

Using Mathematics

- You have 20 minutes to complete this test.
- Calculator not allowed.

1 Complete the missing numbers from the equivalent fractions.

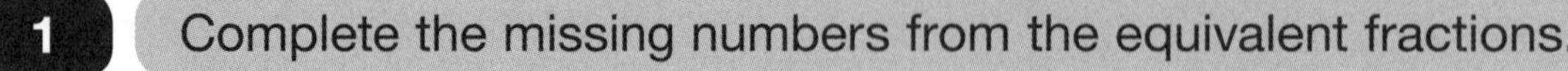

a. $\frac{3}{4} = \frac{6}{\square} = \frac{\square}{12} = \frac{12}{\square}$

1 mark

b. $\frac{5}{8} = \frac{\square}{16} = \frac{15}{\square} = \frac{\square}{32}$

1 mark

2 A number squared and a number cubed both equal 64.

Find the numbers.

$\square^2 = 64 = \square^3$

1 mark

3 Calculate the area of this parallelogram.

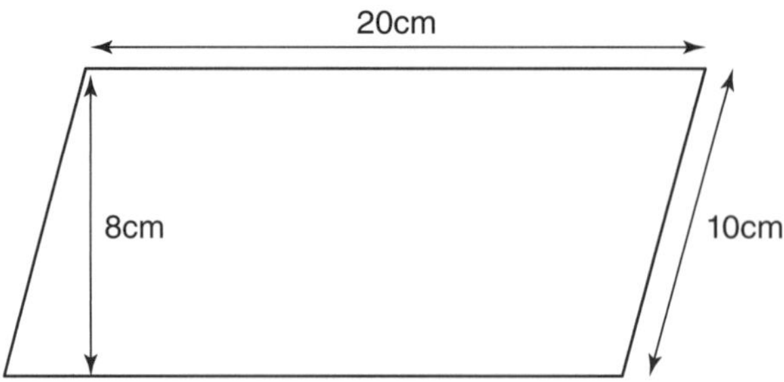

1 mark

4 Sam sells t-shirts from a market stall.

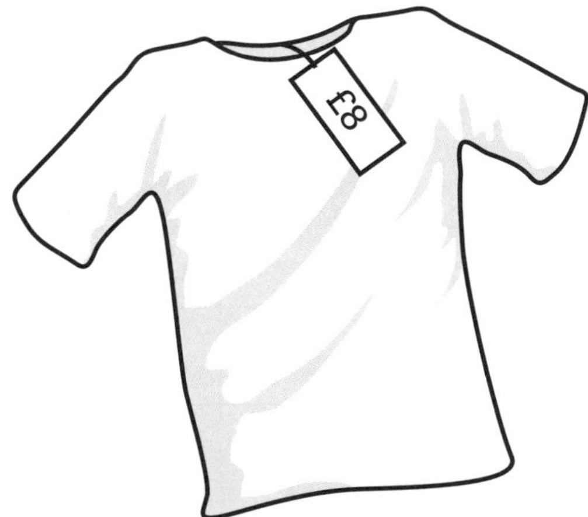

Each t-shirt costs £8

Sam goes to 12 markets.

The mean number of t-shirts sold at the markets is 20

How much did Sam sell all the t-shirts for?

Show your working

£

2 marks

5 This is a map of an island.

It is drawn on a centimetre square grid.

Each square centimetre represents 1 square kilometre.

What is the approximate area of the island?

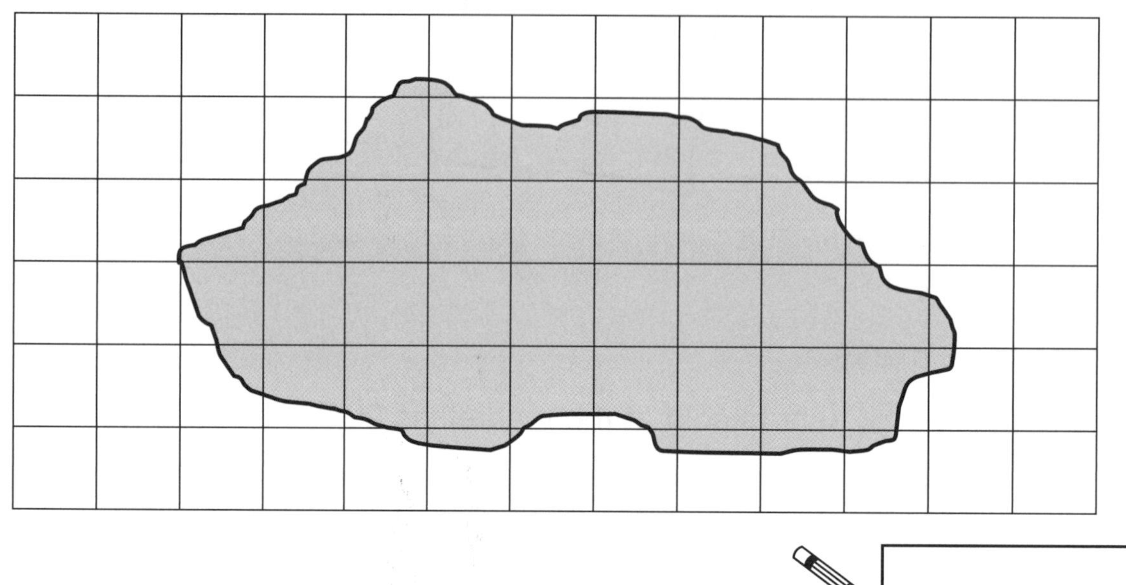

km^2

1 mark

6 This table gives approximate conversions between kilograms and pounds.

kilograms	pounds
1	2.2
2	4.4
4	8.8
8	17.6
16	35.2

Use the approximations to convert 7 kilograms into pounds.

1 mark

7 Calculate the volume of this cuboid.

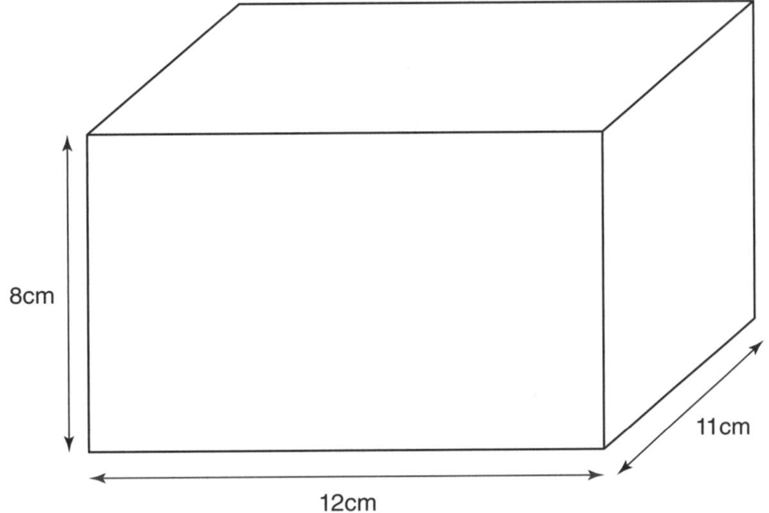

1 mark

8 Calculate the missing numbers.

a. $4\frac{1}{2} + \square = 10\frac{7}{8}$

1 mark

b. $8\frac{11}{12} - \square = 5\frac{1}{12}$

1 mark

9 This sequence decreases in equal steps.

Find the missing number.

6 –3  –17 –26

1 mark

10 The letters stand for numbers.

$$A \times B = 48$$
$$A \div B = 3$$

Which numbers do the letters stand for?

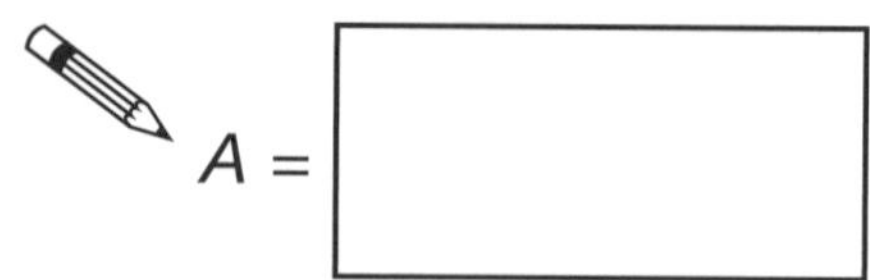

$A =$ ☐

$B =$ ☐

2 marks

11 The points of 3 vertices of a parallelogram are marked on the grid with the symbol ●

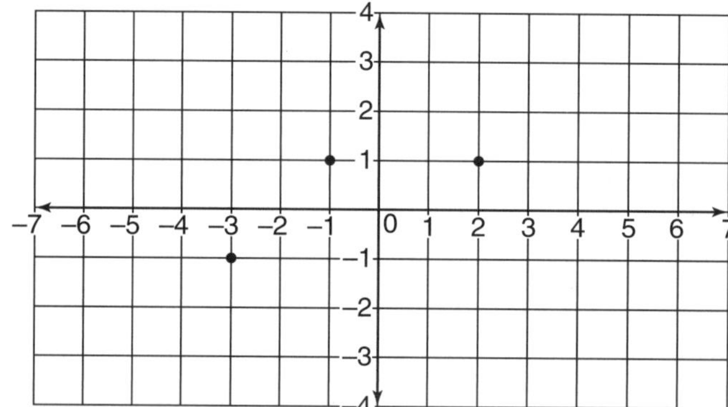

a. Give the coordinates of the 3 marked vertices.

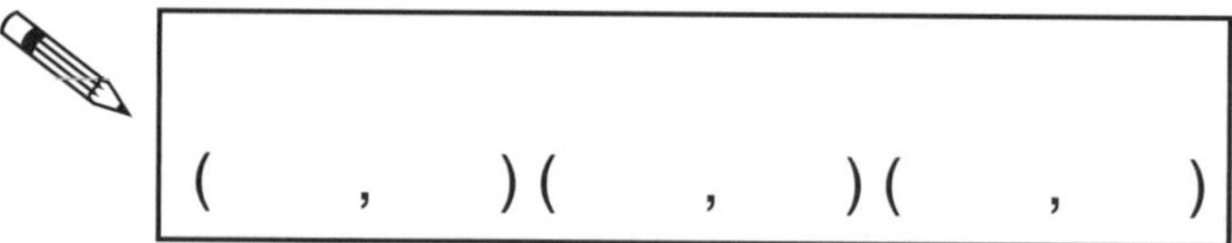

1 mark

Draw lines to complete the parallelogram.

Use a ruler.

b. What could the coordinates of the fourth vertex be?

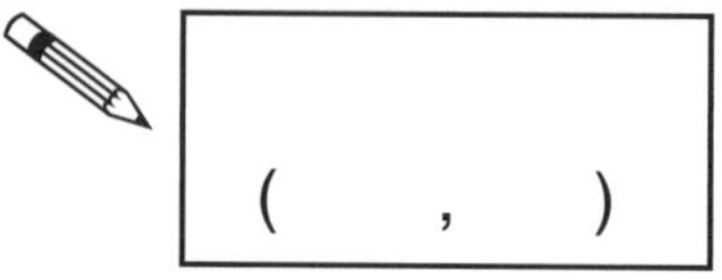

1 mark

Total ______ / 16 marks

Arithmetic: Answers

Test 1

1. 579 **(1 mark)**
2. 8103 **(1 mark)**
3. $1\frac{6}{10}$ (Accept $\frac{16}{10}$) **(1 mark)**
4. 604 **(1 mark)**
5. 84 **(1 mark)**
6. 18 444 **(1 mark)**
7. $\frac{13}{5}$ **(1 mark)**
8. $\frac{5}{12}$ **(1 mark)**
9. 50 **(1 mark)**
10.

		3	2	8
	×		2	6
	1	9	6	8
	6	5	6	0
	8	5	2	8

(2 marks: 1 mark for using long multiplication with no more than one error; 1 mark for correct answer. Do not award any marks if the 0 for multiplying by a ten is missing.)

11.

				7	1
2	5	1	7	7	5
		1	7	5	
				2	5
				2	5
					0

(2 marks: 1 mark for using long division with no more than one error; 1 mark for correct answer)

12. 102 **(1 mark)**
13. –8 **(1 mark)**
14. $\frac{1}{15}$ **(1 mark)**
15. 0.49 **(1 mark)**

Test 2

1. 597 **(1 mark)**
2. 2 **(1 mark)**
3. 3888 **(1 mark)**
4. 1369 **(1 mark)**
5. 1000 **(1 mark)**
6. 2490 **(1 mark)**
7. 3702 **(1 mark)**
8. 670 **(1 mark)**
9. $3\frac{3}{4}$ (Accept $\frac{15}{4}$) **(1 mark)**
10. –17 **(1 mark)**
11. 4480

(2 marks: 1 mark for reaching 560 × 8 (calculation in brackets, 305 – 297, must be completed first); 1 mark for correct answer)

12. 3 **(1 mark)**
13.

		6	4	8
	×		3	5
	3	2	4	0
1	9	4	4	0
2	2	6	8	0

(2 marks: 1 mark for using long multiplication with no more than one error; 1 mark for correct answer. Do not award any marks if the 0 for multiplying by a ten is missing.)

14.

			1	8	1
3	2	5	7	9	2
		3	2		
		2	5	9	
		2	5	6	
				3	2
				3	2
					0

(2 marks: 1 mark for using long division with no more than one error; 1 mark for correct answer)

Test 3

1. 81.5 **(1 mark)**
2. 907 **(1 mark)**
3. 240 **(1 mark)**

4. 200 **(1 mark)**
5. 95 920 **(1 mark)**
6. 5732 **(1 mark)**
7. $1\frac{3}{5}$ (Accept $\frac{8}{5}$) **(1 mark)**
8. $\frac{71}{100}$ **(1 mark)**
9. 113 **(1 mark)**
10.

		5	0	6
	×		5	2
	1	0	1	2
2	5	3	0	0
2	6	3	1	2

(2 marks: 1 mark for using long multiplication with no more than one error; 1 mark for correct answer. Do not award any marks if the 0 for multiplying by a ten is missing.)

11. $3\frac{1}{8}$ **(1 mark)**
12. $\frac{2}{12}$ (Accept $\frac{1}{6}$) **(1 mark)**
13.

			2	4		
3	2	7	6	8		
		6	4			
		1	2	8		
		1	2	8		
				0		

(2 marks: 1 mark for using long division with no more than one error; 1 mark for correct answer.)

14. 30
(2 marks: 1 mark for correct order of calculation (subtraction, division, addition); 1 mark for correct answer.)

Test 4

1. $\frac{7}{12}$ **(1 mark)**
2.

	1	3	8	6				1	6	4	3
+		2	5	7			–		4	0	6
	1	6	4	3				1	2	3	7

(2 marks: 1 mark for a correct first calculation, e.g. 1386 + 257 = 1643; 1 mark for correct answer)

3. 141 **(1 mark)**
4. **80** × 10 + 80 × **6**
(1 mark: both answers needed for 1 mark)
5. 87 **(1 mark)**
6. 846 **(1 mark)**
7. 70 **(1 mark)**
8. $11\frac{19}{20}$ **(1 mark)**
9. $\frac{1}{4}$ (Accept $\frac{6}{24}$) **(1 mark)**
10. 0.36 **(1 mark)**
11. 55 932 **(1 mark)**
12.

		7	1	5
	×		3	8
	5	7	2	0
2	1	4	5	0
2	7	1	7	0

(2 marks: 1 mark for using long multiplication with no more than one error; 1 mark for correct answer. Do not award any marks if the 0 for multiplying by a ten is missing.)

13. 304 060 **(1 mark)**
14. $\frac{1}{24}$ **(1 mark)**
15. 7.2 **(1 mark)**

Test 5

1. 735 254 **(1 mark)**
2. 35 **(1 mark)**
3. 16 181 **(1 mark)**
4. 6 **(1 mark)**
5. 1000 **(1 mark)**
6. $2\frac{2}{5}$ (Accept $\frac{12}{5}$) **(1 mark)**
7. 75% **(1 mark)**
8. 14.25 **(1 mark)**
9. 6.3 **(1 mark)**
10. $\frac{1}{10}$ **(1 mark)**

11.

			2	7	9	.	5
2	8	7	8	2	6	.	0
		5	6				
		2	2	2			
		1	9	6			
			2	6	6		
			2	5	2		
				1	4	0	
				1	4	0	
						0	

(Accept $279\frac{14}{28}$, $279\frac{1}{2}$, 279 r 14)

(2 marks: 1 mark for using long division with no more than one error; 1 mark for correct answer)

12. 10 **(1 mark)**

13.

		5	9	1	8
	×			5	6
	3	5	5	0	8
2	9	5	9	0	0
3	3	1	4	0	8

(2 marks: 1 mark for using long multiplication with no more than one error; 1 mark for correct answer. Do not award any marks if the 0 for multiplying by a ten is missing.)

14. 0.608 **(1 mark)**

15. $5\frac{7}{10}$ **(1 mark)**

Test 6

1. 500 000 **(1 mark)**

2. $\frac{43}{8}$ **(1 mark)**

3. $2\frac{1}{2}$ (Accept $2\frac{3}{6}$, $\frac{15}{6}$) **(1 mark)**

4. 4985 **(1 mark)**

5. 4498 **(1 mark)**

6. 100 000 **(1 mark)**

7. 107.54 **(1 mark)**

8. 0.076 **(1 mark)**

9. $\frac{1}{10}$ (Accept $\frac{2}{20}$) **(1 mark)**

10. $11\frac{11}{20}$ **(1 mark)**

11. 62 **(1 mark)**

12.

		7	5	9	3
	×			7	4
	3	0	3	7	2
5	3	1	5	1	0
5	6	1	8	8	2

(2 marks: 1 mark for using long multiplication with no more than one error; 1 mark for correct answer. Do not award any marks if the 0 for multiplying by a ten is missing.)

13.

				8	6	.	7	5
2	4	2	0	8	2			
		1	9	2				
			1	6	2			
			1	4	4			
				1	8	0		
				1	6	8		
					1	2	0	
					1	2	0	
							0	

(Accept $86\frac{18}{24}$, $86\frac{3}{4}$, 86 r 18)

(2 marks: 1 mark for using long division with no more than one error; 1 mark for correct answer.)

Using Mathematics: Answers

Test 1

1. 80 **(1 mark)**
2. 7.20 (Accept 7.20am, 7.20pm, 19.20, 20 past 7) **(1 mark)**
3. 5272 5277 5727 5772 **(1 mark)**
4. 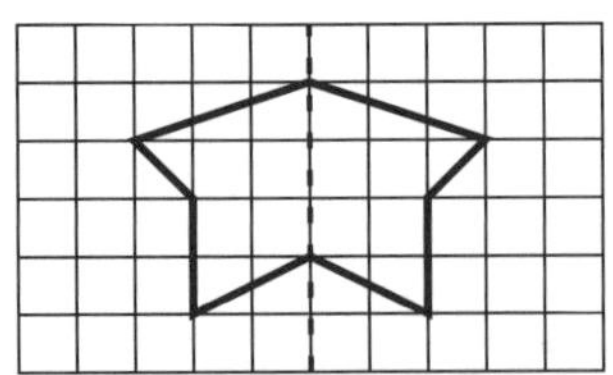

   (Accept lines drawn to within 2mm of vertices. Ignore lines that are not straight.) **(1 mark)**
5. 484 minutes **(1 mark)**
6. 460 305 **(1 mark)**
7. 2 and 29 (Accept answers in either order.) **(1 mark)**
8. $\frac{6}{10} = \frac{3}{5}$

   (2 marks: 1 mark for each correct fraction)
9. 13.25 (Accept 1.25pm, 25 past 1) **(1 mark)**
10. 150 **(1 mark)**
11. 10 and 5 (Accept answers in either order.) **(1 mark)**
12. **a.** A cube has **6** faces. **(1 mark)**
    b. A cube has **12** edges. **(1 mark)**
    c. A cube has **8** vertices. **(1 mark)**

Test 2

1. 493 **(1 mark)**
2. 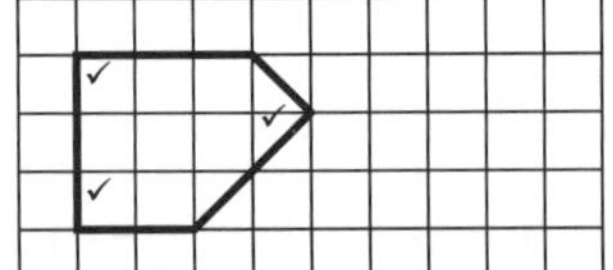

   (2 marks: 2 marks for three angles marked correctly; 1 mark for two correctly marked angles)
3. 5000 **(1 mark)**
4. 1980 **(1 mark)**
5. 3, 6, 15 **(1 mark)**
6. $\frac{7}{10}, \frac{13}{20}, \frac{61}{100}, \frac{3}{5}$ (Accept $\frac{70}{100}, \frac{65}{100}, \frac{61}{100}, \frac{60}{100}$) **(1 mark)**
7. The sides and angles are equal and there are 5 sides. (Accept: The sides and angles in Shape A are equal but they are not in Shape B. Also accept: Shape B is also a pentagon as it has 5 sides, although its sides and/or angles are not all equal.) **(1 mark)**
8. 30cm **(1 mark)**
9. 8°C **(1 mark)**
10. £4.50 (Do not accept £4.5) **(1 mark)**
11. 167, 617, 671, 761

    (2 marks: 1 mark for three correct numbers)
12. 6 **(1 mark)**
13. $\frac{80 - \left(\frac{1}{4} \times 80 + 30\right)}{2} = 15$ children

    (2 marks: 1 mark for correct working, but wrong answer)

Test 3

1. $\frac{5}{8}$ **(1 mark)**
2. 2 × 8 (Accept 16) **(1 mark)**
3. **a. i.** 5.2 **ii.** 5

   (1 mark: Both answers needed for 1 mark.)

   b. i. 14.7 **ii.** 15

   (1 mark: Both answers needed for 1 mark.)
4. Lines completed as shown.

   4.6m — 460cm
   46mm — 4.6cm
   460m — 0.46km
   4.6mm — 0.46cm
   46cm — 0.46m

   (2 marks: 1 mark for two or three correctly drawn lines.)
5. 6300 (90 × 70) **(1 mark)**

6. 540 280 **(1 mark)**

7. 180° **(1 mark)**

8. **a.** 12 boxes (Do not accept 12 with any remainder, decimal or fraction.) **(1 mark)**

   b. 4800 packs **(1 mark)**

9. 4th box ticked only. **(1 mark)**

10. 6cm **(1 mark)**

11. **a.** Possible answers include: 24 × 1, 12 × 2, 8 × 3 (Accept answers with decimals if correct, e.g. 16 × 1.5. Do not accept 6 × 4) **(1 mark)**

    b. Possible answers include: 9 × 1, 8 × 2, 7 × 3, 5 × 5 (Accept answers with decimals if correct, e.g. 8.5 × 1.5. Do not accept 6 × 4) **(1 mark)**

12. $\frac{45}{75}$, $\frac{18}{30}$, $\frac{36}{60}$ circled only.

(2 marks: 1 mark for two fractions correctly circled)

Test 4

1. 2nd and 3rd boxes ticked only.

(2 marks: 1 mark for either box ticked)

2. 3.09, 3.39, 9.09. 9.93, 39.39 **(1 mark)**

3. 1, 36, 64 circled only.

(2 marks: 1 mark for two numbers correctly circled)

4. 55 + 35 + 50 = 130; 130 minutes = 2 hours 20 minutes

(2 marks: 1 mark for correct working, but wrong answer)

5. $\frac{56 - (17 \times 2)}{2} = 11\text{cm}$

(2 marks: 1 mark for correct working, but wrong answer)

6. $\frac{2}{3}$ **(1 mark)**

7. 15.75 **(1 mark)**

8. $x = 70°$ **(1 mark)**

   $y = 40°$ **(1 mark)**

9. 10 × 3 + 0.1 × 50 = £35

(2 marks: 1 mark for correct working, but wrong answer)

Test 5

1. 200 – (40 + 45 + 25 + 30) = 60 minutes

(2 marks: 1 mark for correct working, but wrong answer)

2. £12.04 **(1 mark)**

3. 3rd box ticked only. **(1 mark)**

4. 85% **(1 mark)**

5. **a.** 4512 miles **(1 mark)**

   b. 47 gallons **(1 mark)**

6. 20 ÷ 8 = 2.5, 600 × 2.5 = 1500kg

(2 marks: 1 mark for correct working, but wrong answer)

7. 2, 2

(2 marks: 1 mark for each correct answer)

8. 450ml **(1 mark)**

9. 48 miles **(1 mark)**

10.

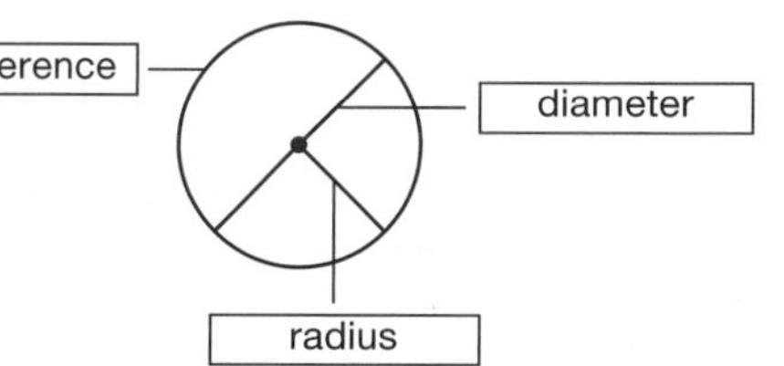

(2 marks: 1 mark for two correct answers)

Test 6

1. 28th (May) **(1 mark)**

2. 18 minutes **(1 mark)**

3. **a.** 115 706 **(1 mark)**

   b. 250 850 **(1 mark)**

4. 2.903, 2.9, 2.593, 2.33 **(1 mark)**

5. **a.** 3 **(1 mark)**

   b. 12 **(1 mark)**

6. 64cm **(1 mark)**

7. 270° **(1 mark)**

8. **a.** 319 500 **(1 mark)**

   b. 300 000 **(1 mark)**

9. 4th box ticked only.

10. □ = 3

    △ = 12 **(1 mark: 1 mark for both correct answers)**

11. **a.** (3,2) marked.

(1 mark: 1 mark if the point (3,2) has been marked or used as a vertex in drawing the square)

    b. (–2, –1) **(1 mark)**

12. $\frac{15}{24}$, $\frac{5}{6}$

(2 marks: 1 mark for each correct answer)

Test 7

1. **a.** 29 games **(1 mark)**
   b. 10 games **(1 mark)**
2. 19, 29, 59
   (2 marks: 1 mark for two correct answers)
3. 6.50 (Accept 6.50am, 6.50pm, 18.50, 10 to 7) **(1 mark)**
4. $2\frac{1}{6}$
   (3 marks: 1 mark for correct addition: $\frac{26}{12}$; 2 marks for correct mixed number: $2\frac{2}{12}$)
5. –5°C **(1 mark)**
6. 52, 46, 34
   (2 marks: 1 mark for any two correct numbers correctly placed in the sequence, e.g. 52, 44, 34)
7. $\frac{1}{2}$, $\frac{7}{12}$, $\frac{2}{3}$, $\frac{3}{4}$, $\frac{5}{6}$
   (Accept $\frac{6}{12}$, $\frac{7}{12}$, $\frac{8}{12}$, $\frac{9}{12}$, $\frac{10}{12}$) **(1 mark)**
8. 6 200 315 **(1 mark)**
9. $x = 100°$ **(1 mark)**
   $y = 80°$ **(1 mark)**
10. **a.** 37.5 **(1 mark)**
    b. 600 **(1 mark)**

Test 8

1. 0.25 litres **(1 mark)**
2. 86 258 – (35 453 + 27 328) = 23 477 saloons
   (2 marks: 1 mark for correct working, but wrong answer)
3. **a.** 272 books **(1 mark)**
   b. 6528 books **(1 mark)**
4. 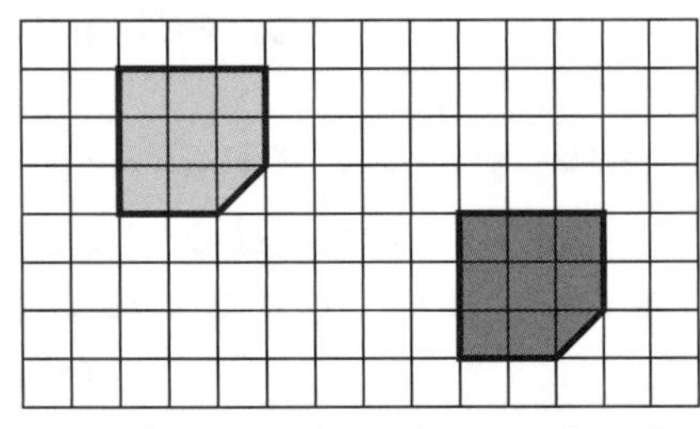
   (2 marks: 2 marks for drawing as shown; 1 mark for correctly orientated and sized shape translated 7 units right or 3 units down)
5. 55° (Accept angle drawn +/– 2°) **(1 mark)**
6. 2 ÷ 5 = ticked only. **(1 mark)**
7. 10 – (2.3 + 1.6) = 6.1kg or 10 000 – (2300 + 1600) = 6100g (Accept 6.1kg or 6100g; Units must be correct, e.g. Do not accept 6.1g or 6100kg.)
   (2 marks: 1 mark for correct working, but wrong answer)
8. 4 **(1 mark)**
9. Pyramid **(1 mark)**
10. (2365 + 3850) ÷ 24 = 258 boxes (Do not accept 258 with any remainder, decimal or fraction.)
    (2 marks: 1 mark for correct working, but wrong answer)

Test 9

1. 192 boxes **(1 mark)**
2. 60, 6
   (1 mark: both answers needed for 1 mark)
3. 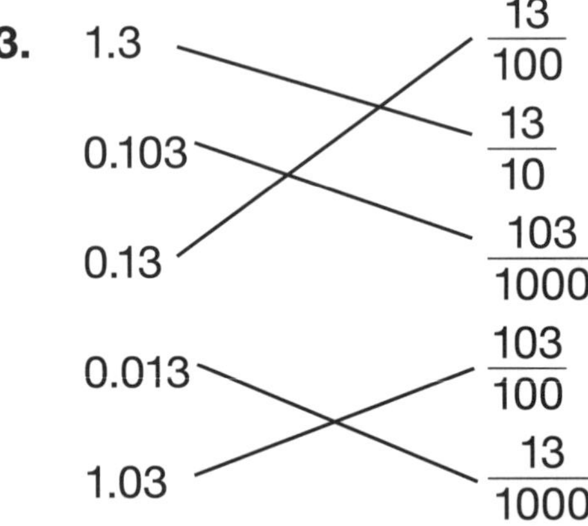

   (2 marks: 1 mark for three or four correctly drawn lines)
4. **a.** 3°C **(1 mark)**
   b. 9°C **(1 mark)**
   c. 3 days **(1 mark)**
5. 8 542 135 **(1 mark)**
6. (9 × 9) – (12 × 6) = 81 – 72 = 9cm^2
   (2 marks: 1 mark for at least one correct calculation of an area i.e. 81 or 72)
7. 13 754 **(1 mark)**
8. **a.** 120 **(1 mark)**

b. 300 **(1 mark)**

c. 25% **(1 mark)**

9. 16 **(1 mark)**

Test 10

1. 59 650 **(1 mark)**

2. 7.324 **(1 mark)**

3. 10, 75, 4

(2 marks: 1 mark for two correct answers)

4.

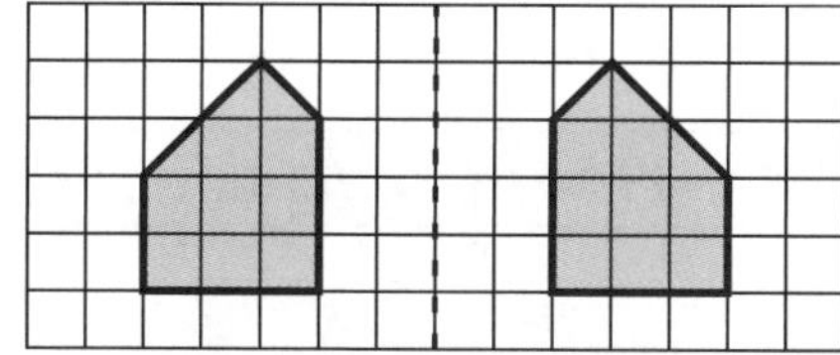

(2 marks: 1 mark for correctly reflected shape in incorrect position)

5. Possible answers:

2×6, 3×4 in either order. (Also accept 1×12 in either order. Do not accept fractions or decimals.) **(1 mark)**

6. **a.** 32 **(1 mark)**

b. 0.047 **(1 mark)**

7. 3rd box ticked only. **(1 mark)**

8. 48, 20 **(1 mark)**

9. $x = 145°$ **(1 mark)**

$y = 35°$ **(1 mark)**

10. **a.** $9cm^2$ **(1 mark)**

b. Square, with 3cm sides (Accept a square with sides of 3cm drawn anywhere on the grid. Accept written answer, e.g. a square with 3cm sides.) **(1 mark)**

Test 11

1. 4th box ticked only. **(1 mark)**

2. 2 634 891 circled only. **(1 mark)**

3. **a.** 2nd box ticked only. **(1 mark)**

b. 3rd box ticked only. **(1 mark)**

4. **(1 mark)**

	6	5	1	8
–	3	0	9	4
	3	4	2	4

5. 20 – (7 + 3.20 + 3.20 + 1.50 + 0.90) = £4.20 (Do not accept £4.2)

(2 marks: 1 mark for correct working, but wrong answer)

6.

(2 marks: 1 mark for each shape correctly orientated and sized)

7. **a.** 7 apples **(1 mark)**

b. 35 apples **(1 mark)**

8. 5 (Accept correct totals all listed: 5, 6, 7, 8, 9) **(1 mark)**

9. 6, 9

(2 marks: 1 mark for each answer)

10. $x = 60°$ **(1 mark)**

$y = 120°$ **(1 mark)**

Test 12

1. **a.** 8, 9, 16

(1 mark: all answers needed for 1 mark)

b. 10, 24, 20

(1 mark: all answers needed for 1 mark)

2. 8, 4 **(1 mark)**

3. $160cm^2$ **(1 mark)**

4. $20 \times 12 \times 8 = £1920$

(2 marks: 1 mark for correct working, but wrong answer)

5. $29km^2$ +/– $1km^2$ **(1 mark)**

6. 15.4 pounds **(1 mark)**

7. $1056cm^2$ **(1 mark)**

8. **a.** $6\frac{3}{8}$ **(1 mark)**

b. $3\frac{10}{12}$ (Accept $3\frac{5}{6}$) **(1 mark)**

9. –12 **(1 mark)**

10. $A = 12$, $B = 4$

(2 marks: 1 mark for numbers in reverse order)

11. **a.** (–3, –1) (–1, 1) (2, 1) **(1 mark)**

b. Possible answers:

(0, –1) (–6, –1) (4, 3) **(1 mark)**

Progress Report

Fill in your total marks for each completed test.

Colour the stars to show how you feel after completing each test.

☆ = needs practice ☆☆ = nearly there ☆☆☆ = got it!

Arithmetic

Test	Marks	How do you feel?
Test 1	/ 17	☆ ☆ ☆
Test 2	/ 17	☆ ☆ ☆
Test 3	/ 17	☆ ☆ ☆
Test 4	/ 17	☆ ☆ ☆
Test 5	/ 17	☆ ☆ ☆
Test 6	/ 15	☆ ☆ ☆

Using Mathematics

Test	Marks	How do you feel?
Test 1	/ 15	☆ ☆ ☆
Test 2	/ 16	☆ ☆ ☆
Test 3	/ 17	☆ ☆ ☆
Test 4	/ 15	☆ ☆ ☆
Test 5	/ 15	☆ ☆ ☆
Test 6	/ 17	☆ ☆ ☆
Test 7	/ 17	☆ ☆ ☆
Test 8	/ 15	☆ ☆ ☆
Test 9	/ 15	☆ ☆ ☆
Test 10	/ 15	☆ ☆ ☆
Test 11	/ 16	☆ ☆ ☆
Test 12	/ 16	☆ ☆ ☆

KS2 Success

Age 7-11

English

Tests

Laura Griffiths

Contents

Reading

TEST 1

- Questions 1–19 are about *A Mad Tea Party* (see pages 66–67 in the reading booklet).
- You have 30 minutes to complete this test.

1 Where is the scene set?

____ 1 mark

2 Why is Alice very small?

____ 1 mark

3 Look at the opening conversation.

Find and **copy** the clause Alice says just before she sits down at the table.

____ 1 mark

4 Which character offered Alice a drink?

Tick **one**.

the Hare ☐

the Hatter ☐

the Narrator ☐

the Dormouse ☐

1 mark

5 Use the following line to answer question 5.

Alice: *Then it wasn't very* ***civil*** *of you to offer it!*

What does the word *civil* mean in this line?

Tick **one**.

polite ☐

angry ☐

old ☐

rude ☐

1 mark

6 Look at the Narrator's lines.

Identify **one** phrase that suggests Alice is correct in saying there is room for her to sit down.

____

____

1 mark

7 Explain why Alice says '*it's very rude*'.

____

____

1 mark

8 What stops the characters arguing?

____

____

1 mark

9 At the end of the passage the Narrator says '*The conversation* ***drops***'.

What does the word *drops* mean in this context?

____ 1 mark

10 After reading the text, how would you describe the character Alice?

____

____

____ 2 marks

11 This text has been written as a play script.

Give **two** features of the text that support this purpose.

1. ____

2. ____ 2 marks

12 *(The Hatter takes his watch out of his pocket and looks at it uneasily; he shakes it and holds it to his ear. He then dips it in his cup of tea and looks at it again.)*

(a) What adverb gives more information about the Hatter's personality?

____ 1 mark

(b) Describe the Hatter's character throughout the scene.

____

____

____ 1 mark

13 Look at the title of the extract: ***A Mad Tea Party***.

Explain fully why this is an appropriate title, referring to the text in your answer.

____

____

________________________________ 3 marks

14 **Find** and **copy one** stage direction from the text.

________________________________ 1 mark

15 Whose hair did the Hatter say needs cutting?

________________________________ 1 mark

16 What is a synonym for the word *riddle*?

Tick **one**.

solution ☐

puzzle ☐

sentence ☐

argument ☐

1 mark

17 At the beginning of the extract Alice says '*Hello, mind if I join you* ***all****?*'

Who does she mean by *all*?

____ 1 mark

18 Explain the role of the Narrator in the play script.

____

____ 2 marks

19 Look at the title of the extract: ***A Mad Tea Party***.

Which of the following would be the most suitable replacement for this title?

Tick **one**.

An Ordinary Tea Party ☐

An Unusual Tea Party ☐

A Boring Tea Party ☐

A Quiet Tea Party ☐

1 mark

Total ________ / 25 marks

TEST 2

Reading

- Questions 1–15 are about *James and the Giant Peach* (see pages 68–70 in the reading booklet).
- You have 30 minutes to complete this test.

1 Look at the sentence.

It felt soft and warm and slightly furry, like the skin of a baby mouse.

Tick the correct answer.

The sentence contains:

a metaphor ☐

a simile ☐

personification ☐

alliteration ☐

1 mark

2 *James stopped and stared at the speakers, his face white with horror. He started to stand up, but his knees were shaking so much he had to sit down again on the floor.*

Write how James is feeling.

____

1 mark

3 *The creatures, some sitting on chairs, others reclining on a sofa, were all watching him* ***intently****.*

Which word is the correct synonym of *intently*?

Tick **one**.

casually ☐

closely ☐

quickly ☐

strangely ☐

1 mark

4 Draw a line to match each creature to the description James gives when he **first** meets them.

Old-Green-Grasshopper	nine black spots on her scarlet shell
Spider	enormous
Silkworm	as large as a large dog
Ladybird	thick and white

1 mark

5 *He glanced behind him, thinking he could* ***bolt*** *back into the tunnel . . .*

What does the word *bolt* mean in this sentence?

____

1 mark

6 What reason does the Ladybird give for James looking *'...as though he's going to faint any second'*?

____

____

1 mark

7 Explain as fully as you can what the Ladybird means when she tells James *'You are one of us now, didn't you know that?'*

____

____

1 mark

8 Put a tick in the correct box to show whether each of the following statements is **true** or **false**.

	True	**False**
A fence surrounded the peach.		
The tunnel was dry and cold.		
James bumped his head on the stone in the middle of the peach.		
James couldn't stand up because the peach wasn't big enough.		
The creatures were at least the same size as James.		

1 mark

9 What **effect** is the author trying to create in the sentence below?

Four pairs of round black glassy eyes were all fixed upon James.

____

____

____

____________________ 2 marks

10 Which words does the author use to create a magical setting in the first paragraph?

Use examples from the text below.

The garden lay soft and silver in the moonlight. The grass was wet with dew and a million dewdrops were sparkling and twinkling like diamonds around his feet. And now suddenly, the whole place, the whole garden seemed to be alive with magic.

____

____

____________________ 2 marks

11 Explain why this statement by the Centipede is funny.

And meanwhile I wish you'd come over here and give me a hand with these boots. It takes me hours to get them all off by myself.

____

____

____________________ 1 mark

12 Which **two** parts of his body did James put into the peach first?

1. ______________________________

2. ______________________________ 1 mark

13 *The floor was **soggy** under his knees.*

What does the word *soggy* mean in this sentence?

______________________________ 1 mark

14 When James realised that the hole in the peach was a tunnel, how did he feel?

______________________________ 1 mark

15 When James first entered the room at the centre of the peach, which **two** characters were seated next to the Spider?

1. ______________________________

2. ______________________________ 1 mark

Total ________/17 marks

Reading

TEST 3

- Questions 1–16 are about the Isles of Scilly (see pages 71–73 in the reading booklet).
- You have 30 minutes to complete this test.

1 What is the name of the largest island of the Isles of Scilly?

____

1 mark

2 *The Isles of Scilly are made up of five* ***inhabited*** *and approximately 140 other islands.*

What does the word *inhabited* mean?

Tick **one**.

deserted ☐

far away ☐

people live there ☐

tropical ☐

1 mark

3 Look at the paragraph beginning: *To reach the Isles of Scilly from the mainland . . .*

Give **two** ways visitors can reach the islands.

1. ______________________________

2. ______________________________

1 mark

4 Read the advantages and disadvantages of the two methods of transport. Which would you recommend for the following visitors? Give reasons for your answers **using the information in the text**.

A day visitor with no luggage: ______________________________

____

____

______________________________ 2 marks

A family who are staying for a week who want to see dolphins swimming:

____

____

______________________________ 2 marks

5 In the paragraph about plants and animals, explain what is meant by the following:

a large natural greenhouse

____

______________________________ 1 mark

6 What does the word *common* mean in the sentence below?

There are many different types of birds on the islands, ranging from ***common*** *birds such as sparrows, thrushes and gulls, to more unusual birds like puffins, cuckoos and wheatears.*

____

______________________________ 1 mark

7 When describing what tourists can see and do on the Isles of Scilly, the writer has deliberately chosen language that will have an effect on the reader.

Some of the words in the table below are in **bold**.

Explain the effect of these words in each sentence.

Language used	Explanation of the effect of the language
. . . **flock** to race.	
. . . **waiting** to be explored . . .	
. . . **swirling golden** beaches . . .	

3 marks

8 **Find** and **copy** a reason to support each statement in the table below.

Statement	Reason
Very few people own cars.	
The islands are very windy.	
A variety of rare plants and flowers grow.	
Birds migrate to the Isles of Scilly.	

4 marks

9 Would you like to visit the Isles of Scilly? Give **three** reasons for your view. Use the text to help you.

1. ______________________________

2. ______________________________

3. ______________________________

1 mark

10 Compare how someone participating in and someone watching a gig race would be involved.

Participating: ______________________________

____

Watching: ______________________________

____

2 marks

11 Write **one** question you would like to ask someone who lives on the Isles of Scilly.

____

____

1 mark

12 This text was written to inform people about the Isles of Scilly.

Give **two** features of the text that support this purpose.

- ______________________________
- ______________________________

2 marks

13 List **three** activities that the leaflet suggests for tourists to do when visiting the Isles of Scilly.

1. ______________________________

2. ______________________________

3. ______________________________

2 marks

14 What is the name of the main town on St Mary's?

____

1 mark

15 How do the locals usually travel around the islands? Give **two** modes of transport.

1. ______________________________

2. ______________________________

2 marks

16 In what format are you most likely to see this text?

Tick **one**.

a poster ☐

a travel leaflet ☐

a storybook ☐

a school magazine ☐

1 mark

Total ______ /28 marks

TEST 4

Reading

- Questions 1–13 are about About Floella (see pages 74–75 in the reading booklet).
- You have 30 minutes to complete this test.

1 What type of writing is featured in this account?

Tick **one**.

diary ☐

autobiography ☐

biography ☐

story ☐

1 mark

2 Where was Floella born?

____

1 mark

3 What does the word *challenging* mean in the sentence below?

*Initially she found it very different to her life in Trinidad and growing up in two cultures was **challenging**.*

____

____

1 mark

4 Floella Benjamin's family are special to her.

Find and **copy two** sentences that infer Floella had a good relationship with her parents.

Mum: ______________________________

____

Dad: ______________________________

______________________________ 2 marks

5 **(a)** What is the name of Floella's autobiography?

______________________________ 1 mark

(b) Why do you think Floella chose this title? Give an example from the text.

____

____

______________________________ 1 mark

6 What was the name of the drama in which Floella made her TV debut?

______________________________ 1 mark

7 Find the year when each of the events happened in Floella Benjamin's life. Write the answers in the grid below.

Event	Date
Born	
Came to England	
Made her TV debut	
Started her own television production company	
Wrote *Coming to England*	
Trekked the Great Wall of China	

2 marks

8 In what order is the text written?

Tick **one**.

importance ☐

no particular order ☐

chronological ☐

1 mark

9 For which charity does Floella raise money by running the London Marathon?

Tick **one**.

Sickle Cell Society ☐

Action for Children ☐

Barnardo's ☐

NSPCC ☐

1 mark

10 This text was written to inform people about the life of Floella Benjamin.

Give **two** features of the text that support this purpose.

1. ______________________________

2. ______________________________

2 marks

11 Summarise **three** ways we know that Floella has helped children's charities.

1. ______________________________

____

2. ______________________________

____

3. ______________________________

____

3 marks

12 Put a tick in the correct box to show whether each of the following statements about Floella are **fact** or **opinion**.

	Fact	**Opinion**
She worked in a bank.		
She loved her job on *Playschool*.		
She presented *Playschool* for 12 years.		
She has completed ten London Marathons.		
She is a keen runner.		

1 mark

13 Explain how Floella Benjamin's childhood influenced her to help young people across the world.

____

____

____

3 marks

Total ______/**21 marks**

Reading

TEST 5

- Questions 1–15 are about *Street Child* (see pages 76–78 in the reading booklet).
- You have 30 minutes to complete this test.

1 What is the setting for the passage?

____ 1 mark

2 Why did Joseph say *'It's the whole world, this place is'*?

Tick **one**.

He thinks it's the best place to be. ☐

He doesn't know life anywhere else. ☐

1 mark

3 *. . . Jim was looking up at the high walls that surrounded the workhouse, and at the bleak sky above it.*

(a) Give **one** reason why the walls were high.

____ 1 mark

(b) What do you think Jim was thinking as he looked up?

____

____ 1 mark

4 Look at the paragraph that starts *It was impossible to tell . . .*

Explain how in this paragraph Jim realises he's been in the workhouse for a year.

____

____

____

_______________ 2 marks

5 *It was then that the* ***little secret promise*** *that had nestled inside him began to flutter into life like a wild thing.*

At this point in the story, what do you think Jim was thinking? What was the *little secret promise*?

____

____

_______________ 2 marks

6 *The teacher* ***hauled*** *him off his stool . . .*

Which word below is a synonym for the word *hauled*?

Tick **one**.

poked ☐

pushed ☐

persuaded ☐

pulled ☐

1 mark

7 Why did Jim say he didn't mind when Mr Barrack was beating him?

____

____

1 mark

8 What reasons did Tip give for not going with Jim?

____

____

2 marks

9 When describing Jim's feelings throughout the extract, the writer has deliberately chosen language that will have an effect on the reader.

Some of the words in the table below are in **bold**.

Explain the effect of these words in each sentence.

Language used	Explanation of the effect of the language
Jim's wild thoughts **drummed** inside him . . .	
. . . the **beating** inside him was like a wild bird now . . .	
. . . a shimmer of pain and **thrumming** wings.	

3 marks

10 *'A daft boy, you are,' said Tip.*

Do you think Tip was correct? Give examples from the text to support your answer.

2 marks

11 Describe the relationship between the two main characters, Jim and Tip, throughout the extract. Give examples from the text to support your answer.

2 marks

12 Why did Jim say '*Seems like I was born here*'?

1 mark

13 Draw a line to match each event below to show the correct order from 1–6, as it appears in the extract.

Event	Order
Mr Sissons asked for big boys to help.	1
Jim asked Tip to run away with him.	2
Jim tried to remember what life outside the workhouse was like.	3
Tip shared his breakfast with Jim.	4
Mr Barrack gave Jim 'a beating'.	5
Jim decided to climb over the wall and escape.	6

3 marks

14 Underline **two** words which suggest Mr Barrack was pleased he had caught Jim talking.

Mr Barrack sprang down from his chair, his eyes alight with anger and joy.

'You spoke!' he said to Jim, triumphant. 'It was you.'

1 mark

15 While Mr Barrack was beating Jim, what reason did he give for tying a handkerchief under Jim's chin?

____

____

1 mark

Total ________ / 24 marks

TEST
6

Reading

- Questions 1–17 are about *Macavity* (see pages 79–80 in the reading booklet).
- You have 30 minutes to complete this test.

1 What type of animal is Macavity?

____

1 mark

2 What happens when someone goes to investigate every crime Macavity has committed?

____

1 mark

3 Look at the third verse.

This verse tells readers more information about Macavity's:

Tick **one**.

friends ☐

personality ☐

hobbies ☐

appearance ☐

1 mark

4 What does the word **neglect** mean in the third verse?

____

____

1 mark

5 Look at the first line of the poem.

Macavity's a ***Mystery Cat****:*

Explain why the poet describes Macavity as a *Mystery Cat*.

1 mark

6 Why do you think the poet uses **repetition** of the phrase '*Macavity's not there*'?

1 mark

7 Look at verse five.

Explain what Macavity might have done for each crime.

Crime	What do you think Macavity did?
The larder's looted.	
The jewel-case is rifled.	
The milk is missing.	
The greenhouse glass is broken.	

4 marks

8 Use the following line to answer questions 8 (a) and (b).

He always has an alibi, and one or two to spare:

(a) Explain what the line means.

____

____ 1 mark

(b) What does the line tell readers about Macavity's crimes?

____ 1 mark

9 Write **two** questions you would like to ask Macavity about his behaviour.

1. ____

____

2. ____

____ 2 marks

10 What type of poem is *Macavity*?

Tick **one**.

limerick	☐
narrative	☐
Haiku	☐
sonnet	☐

1 mark

11 Describe the poet's use of **rhyme** throughout the poem.

___

___

___

2 marks

12 *MACAVITY WASN'T THERE!*

In the final verse, why do you think the poet changed the repetition and used capital letters?

___

___

___

1 mark

13 Put a tick in the correct box to show whether each of the following statements are **fact** or **opinion**.

	Fact	**Opinion**
It is fun watching Macavity cause trouble.		
Macavity's a ginger cat.		
His coat is dusty.		
Macavity has no-one to care for him.		

1 mark

14 **Find** and **copy** a simile from verse three.

____ 1 mark

15 In verse five, **find** and **copy** **one** word which means the same as *stolen*.

____ 1 mark

16 In the final verse, explain what the poet means by '*Just controls their operations: the Napoleon of Crime!*'

____

____ 2 marks

17 Summarise **three** main ideas the poet implies about Macavity's characteristics throughout the poem.

1. ____

____

2. ____

____

3. ____

____ 3 marks

Total ________/26 marks

Grammar and Punctuation

TEST 1

- You have 30 minutes to complete this test.

1 Tick the boxes where **articles** have been used correctly.

Tick **two**.

an elephant ☐

an drum ☐

a octopus ☐

a superhero ☐

1 mark

2 Tick the word that means ‘to find out’.

Tick **one**.

choose ☐

discover ☐

invite ☐

repeat ☐

1 mark

3 Circle the word that makes most sense in each set of brackets.

We (was / were) going to play outside on our bikes.

I (done / did) well in my maths test this morning.

2 marks

4 Add the **inverted commas** to the direct speech below.

Wait for me! shouted Orla.

Ouch! My hand hurts.

Freddie whispered, Are you scared?

3 marks

5 Circle the **verb** in the sentence below.

Olivia and Sandip were building sandcastles at the beach.

1 mark

6 Which sentence shows the correct agreement between **subject** and **verb**?

Tick **one**.

The authors writed letters to the newspaper.	☐
The authors wrote letters to the newspaper.	☐
The authors write letter to the newspaper.	☐
The authors writes letters to the newspaper.	☐

1 mark

7 Add the missing **full stops** and **capital letters** to the sentence below.

nottingham is located in the east midlands the river that runs through

nottingham is called the river trent

2 marks

8 Circle the **conjunction** in the sentence below.

Since it was very cold outside, Sam decided to fasten his coat.

1 mark

9 Draw a line to match each **prefix** to the root word to make a new word.

dis		haul
mis		honest
over		understood

1 mark

10 Add an **apostrophe** in the correct places to show possession.

Emmas cat had five kittens last night.

The boys changing rooms were locked.

Jamies lunchbox was left overnight in his classroom.

3 marks

11 Add **two dashes** to make the sentence below correct.

William Shakespeare a famous author wrote the play 'Macbeth'.

1 mark

12 Underline the **main clause** in the sentence below.

Although she is younger than me, my sister is much taller.

1 mark

13 The incomplete sentences below are instructions in a recipe.

Add **two adverbials of time** to make the two sentences correct.

______________________ check you have the correct ingredients.

______________________ turn the oven on to 180 degrees.

2 marks

14 Choose **one** of the **question tags** below to complete the sentence.

haven't you	haven't we	didn't you

You've been learning about materials in science, ____________________?

1 mark

15 Chose the correct form of the **past tense** of the verbs below to show a **continuous action**.

to dance	to play

While the band ____________________, I

____________________ with my friends.

1 mark

16 Tick the option which correctly introduces the **subordinate clause** in the sentence below.

The teacher was pleased with the children's work,

____________________ he gave them extra playtime!

Tick **one**.

despite ☐

therefore ☐

however ☐

finally ☐

1 mark

17 Insert a **comma** in the correct place in the sentence below.

Feeling confident the pianist played to a room full of people.

1 mark

18 Write the name of the punctuation mark circled in the sentence below.

The dogs barked(:)they were scared.

____

1 mark

19 In the sentence below, what **word class** is the word **they**? Put a tick next to your answer.

I wish they would be quiet now!

Tick **one**.

adjective ☐

preposition ☐

verb ☐

pronoun ☐

1 mark

20 Which of the events below is the **most likely** to happen?

Tick **one**.

I will wash the car today. ☐

I should go to work. ☐

I might watch a film. ☐

I could play in the garden. ☐

1 mark

Total ________/27 marks

TEST 2

Grammar and Punctuation

- You have 30 minutes to complete this test.

1 Draw a line to match the words to the correct sentence type.

Ouch!	question
How old are you?	statement
The music was very loud.	exclamation

1 mark

2 Circle the **verb** in this sentence.

Dad slowly walked to the shops.

1 mark

3 Read the sentence below.

Circle the most suitable **pronoun** to complete the sentence.

he	his	it	me

Zane ate his dinner and then ______________________ went outside to play.

1 mark

4 Underline the **adverbial** in the sentence below.

Later that evening, we said goodbye and began our journey home.

1 mark

5 Tick the correct word below to complete the following sentence.

The school was closed ______________________ of a fire.

Tick **one**.

although ☐

because ☐

during ☐

after ☐

1 mark

6 Add **three commas** in the correct places in the sentence below.

Mangoes kiwis apples pineapples and strawberries are all types of fruit.

1 mark

7 Write this sentence in the **past tense**.

We laugh at each other's funny jokes.

____

1 mark

8 Circle the **conjunction** in the sentence below.

I am allowed to watch television while eating my dinner.

1 mark

9 Use the words below to complete the table.

ancient	new	small
large	wealthy	poor

	Synonym	Antonym
sad	unhappy	happy
rich		
big		
old		

3 marks

10 Complete the sentences below with the correct **articles**.

The water park was ______________________ amazing sight. It had ______________________ wave machine and ______________________ slide and ______________________ big pool had lots of toys.

1 mark

11 Insert the missing **punctuation** in the sentence below.

The skateboard park located behind the playground is to be used by children who are over eight

1 mark

12 Rewrite the sentence putting **ellipses** in the correct places.

On your marks ready steady go!

____

1 mark

13 Rewrite the sentence below in the **active** voice.

The ancient ruins were visited by the historians.

The historians ____

1 mark

14 Change the **nouns** to **verbs**.

Noun	Verb
simplification	to
suffocation	to
magnification	to

3 marks

15 Circle the **preposition** in the sentence below.

The girl walked up the stairs.

1 mark

16 Rewrite the sentence below, adding a **subordinate clause.**

Remember to use the correct punctuation.

The farmer went into his field.

____

____

____ 2 marks

17 Write a sentence using a **modal verb.**

____ 1 mark

18 Underline the **relative clause** in the sentence below.

The old lady who was shouting at her neighbour was feeling angry. 1 mark

19 Circle the word or words in the sentence below that make it a **question.**

'You're eating all your dinner, aren't you?' 1 mark

Total _________/24 marks

Grammar and Punctuation

TEST 3

- You have 30 minutes to complete this test.

1 Circle **all** the **adverbs** in the sentence below.

Remember to walk down the stairs quietly and carefully shut the door.

2 marks

2 In each sentence circle the letter that should be a **capital letter**.

I have swimming lessons on a wednesday.

My best friend is called lydia.

we went to the opticians to have our eyes checked.

Tenerife is a spanish island.

1 mark

3 Choose **one** of the following **prefixes** to make this word correct.

super	anti	dis	auto

___________clockwise

1 mark

4 Circle the **two** words that show a **command** in the sentences below.

Put the flour and the butter in the bowl. Mix them together carefully.

1 mark

5 Write a correct **pronoun** in the space to make the sentence correct.

The boy ran up the school drive. ______________________ was late again!

1 mark

6 Finish off the sentence below using these words.

Use commas in the correct places.

milk	eggs	cheese	apples	and

At the supermarket I bought ______________________________________

____.

1 mark

7 Complete the sentence below with a **contraction** that makes sense.

Why ______________________________ you find your homework?

1 mark

8 Choose the correct form of the **past tense** of the verbs below to show a **continuous action**.

When I ______________________________ (to dance) in the show I

______________________________ (to smile) all the time.

2 marks

9 Add **inverted commas** to punctuate the speech below.

What time does the train leave the station? Mary asked the guard.

10:03, the guard answered.

2 marks

10 Rewrite the sentence below changing the verb to the **present tense**.

The bird has gone to look for food.

____

1 mark

11 Where should a **question mark** be added to make the sentence below correct?

Tick **one**.

'Where are you going' the bus driver asked.

↑ ☐ ↑ ☐ ↑ ☐

1 mark

12 Use the **conjunctions** in the box to correctly complete the sentence below.

Use each conjunction **once**.

or	but	and

The books ______________ DVDs need returning to the library on Monday ______________ Tuesday ______________ remember it is closed each day for lunch.

1 mark

13 Insert a **comma** in the correct place in the sentence below.

Feeling a little anxious the children walked into the hall.

1 mark

14 Circle **two determiners** in the sentence below.

The weather forecast is good for every day next week.

1 mark

15 Rewrite the sentence below so that it starts with a **subordinate clause.**

Remember to use a **comma** in the correct place.

I read a book while I was waiting to see the doctor.

____

2 marks

16 The sentence below has been written in the passive voice.

Rewrite it in the **active** voice.

The ball was chewed by the dog.

____

1 mark

17 Fill in the correct verb form to complete the sentence below so that it becomes more **formal**.

If the bus ____________________ late again, the children would be cross.

1 mark

18 Which of the sentences uses **dashes** correctly?

Tick **one**.

The cat – fast asleep – on the rug – was keeping warm by the fire. ☐

The cat – fast asleep on the rug – was keeping warm by the fire. ☐

The cat fast asleep, on the rug – was keeping warm – by the fire. ☐

The cat – fast asleep on the rug, – was keeping warm by the fire. ☐

1 mark

19 Underline the **fronted adverbial** in the sentence below.

With his sports kit on, Tom was ready for the game.

1 mark

20 Rewrite the sentence below adding **ellipses** correctly.

There was a crash a loud bang then darkness!

____

1 mark

21 The sentence below has been written in the active voice.

Rewrite it in the **passive** voice.

The cat knocked over the vase.

____

1 mark

Total ________/25 marks

TEST 4

Grammar and Punctuation

- You have 30 minutes to complete this test.

1 Circle **all** the **adjectives** in the sentence below.

The creepy shadows darted quickly past the wooden fence, along the hidden path and into the gloomy garage.

1 mark

2 Which word completes the sentence below?

You cannot go on the field today ______________ it is too muddy.

Tick **one**.

however	☐
although	☐
because	☐
and	☐

1 mark

3 Circle the correct word in each set of brackets.

I (did / done) a detailed piece of writing.

They (has / have) been playing outside.

2 marks

4 Rewrite the sentences below adding **apostrophes** in the correct place **to mark possession.**

The dogs tail was wagging.

____

The boys coats were on the floor.

____

The womens changing rooms were busy.

____ 3 marks

5 Draw a line to match each **punctuation mark** to the correct type of sentence.

question	!
exclamation	?
statement	.

1 mark

6 Rewrite the sentence below so that it begins with the **adverbial**.

Use only the same words and remember to punctuate your answer correctly.

Kenzie blew out the candles after the singing.

____

____ 1 mark

7 Add an adjective before each noun to make an **expanded noun phrase**.

The ______________________________ skateboard park.

The ______________________________ holiday.

2 marks

8 Explain how the **comma** changes the meaning in the two sentences below.

I want to eat James.

I want to eat, James.

____

____

____

1 mark

9 Write the name of the punctuation circled in the sentence below.

Then, all of a sudden, lightning struck . . . boom!

____

1 mark

10 Insert a **semicolon** in the most appropriate place in the sentence below.

Grace bought a bouncy ball she played with it in the garden.

1 mark

11 Put a tick in each row to show the type of **noun** highlighted in **bold** in each sentence.

Sentence	Proper noun	Common noun
My birthday is in **January**.		
The **bees** landed on the flower.		
The car towed the **caravan**.		
His brother's name is **Andrew**.		

1 mark

12 Write a suitable **modal verb** in the sentence below.

They really ______________________________ take more care.

1 mark

13 Change the verb below into a **noun**.

'to suffocate' ______________________

verb — noun

1 mark

14 Draw a line to show whether each sentence is written in the **active** or **passive** tense.

Rosie climbed the stairs.	passive
The stairs were climbed by Rosie.	active

1 mark

15 Circle the **conjunction** in the sentence below.

The girls were tired but it was too early to go to bed.

1 mark

16 Circle the word or words in the sentence below that make it a **question.**

'You were expecting this letter today, weren't you?'

1 mark

17 Underline the **relative clause** in the sentence below.

The ice cream van that is in the car park is very popular today.

1 mark

18 Use each preposition from the box to complete the sentences below.

under	in	along

She walked ________________ the canal path.

The dog hid ________________ the table.

I was lost ________________ the maze.

3 marks

Total ________/24 marks

Grammar and Punctuation

TEST 5

- You have 30 minutes to complete this test.

1 Tick **one** box in each row to show whether each word is **singular** or **plural**.

	Singular	**Plural**
children		
dog		
women		
bananas		

1 mark

2 Underline the **adjectives** in this sentence.

The delicious juicy apple fell from the large tree.

1 mark

3 Which sentence contains **two verbs**?

Tick **one**.

I ran home very quickly. ☐

The teacher read a very long story. ☐

The little boy watched TV and ate a sandwich. ☐

The spider crawled underneath the doormat. ☐

1 mark

4 Circle **all** the **determiners** in the sentence below.

An alligator is a large reptile.

1 mark

5 Complete the sentence using suitable **pronouns**.

Rosie's homework was too hard for ____________________ and ____________________ felt sad that ____________________ couldn't do it.

3 marks

6 Which of the events in the sentences below is the **most** likely to happen?

	Tick **one**.
We might go to the cinema tonight.	☐
He will go to school tomorrow.	☐
They could go camping at the weekend.	☐
I can go to your party tomorrow.	☐

1 mark

7 Write a suitable **question tag** at the end of the statements below.

It's a nice day today, ______________________________

You like reading, ______________________________

2 marks

8 Insert a set of **brackets** so that the sentence below is punctuated correctly.

Mrs Jones the Year 5 teacher played the piano in school today.

1 mark

9 Add two **subordinate clauses** to the main clause below.

_______________________________, I went

to the park, _______________________________.

1 mark

10 Insert a **comma** in the correct place in the sentence below.

Hanging upside down the bat made loud noises as the night sky

grew darker.

1 mark

11 Rewrite the sentence below in the **active** form.

The leftover food was eaten by the dog.

____

1 mark

12 Complete the table by inserting a **synonym** and an **antonym**.

Word	Synonym	Antonym
rich		
angry		

2 marks

13 Circle the correct form of the **verb** in each set of brackets.

The tree (sway / sways) gently in the breeze.

The goats (have / has) two horns each.

The children (are / is) too noisy!

1 mark

14 To make a pop-up book, Louis needs three pieces of equipment: scissors, card and glue.

Rewrite these instructions using **bullet points**.

To make a pop-up book you will need three things:

- ____________________
- ____________________
- ____________________

1 mark

15 Choose a suitable **prefix** for the following:

__________cover

__________elect

__________allow

3 marks

16 Write **two conjunctions** to complete the sentence below.

I like drawing ____________________ painting

____________________ my sister prefers writing.

1 mark

17 Rewrite the sentence below so that it begins with an **adverbial**.

Remember to punctuate your sentence correctly.

We went to the cinema after tea.

____

____ 2 marks

18 Explain how the **comma** changes the meaning in the two sentences below.

Jump over, Ben!

Jump over Ben!

____

____ 1 mark

19 Write this as a sentence using **direct speech**.

Remember to punctuate it correctly.

Mrs Shepherd told her class to be quiet before they went into assembly.

____

____ 1 mark

Total ________/26 marks

TEST 6

Grammar and Punctuation

- You have 30 minutes to complete this test.

1 Circle the **conjunction** in the sentence below.

Isla listened carefully so she understood the instructions.

1 mark

2 Put **commas** in the correct places to separate items in the list below.

The milkman delivers fresh produce on a Monday Tuesday Thursday and Saturday.

1 mark

3 Add the pronouns **I** and **me** to the sentences below to make them correct.

Noah and ______________________________ went for a walk in the woods.

'You pushed ______________________________ into the mud.'

1 mark

4 Circle the **preposition** in the sentence.

The cat slept in the warm, cosy basket.

1 mark

5 Which of the sentences below uses the **semi-colon** correctly?

Tick **one**.

Ella played the piano; Alex played the flute. ☐

Ella played; the piano Alex played; the flute. ☐

Ella played; the piano; Alex played the flute. ☐

Ella; played the piano Alex; played the flute. ☐

1 mark

6 Put a tick in each row to show if each word is an **adverb of time** or **place**.

Adverb	Time (when)	Place (where)
frequently		
regularly		
often		
into the bin		

1 mark

7 Underline the **relative clause** in the sentence below.

Brighton, which is a seaside town, is located on the south coast of England.

1 mark

8 Which sentence uses **inverted commas** correctly?

Tick **one**.

It's 'my turn next,' I told my older brother. ☐

'It's my turn next I told,' my older brother. ☐

'It's my turn next, I told my older brother.' ☐

'It's my turn next,' I told my older brother. ☐

1 mark

9 Write a sentence that uses brackets for **parenthesis**.

____

____

1 mark

10 Explain how the placing of a **comma** changes the meaning of the sentences below.

When they met Grandma, Tom and Joseph went swimming.

When they met, Grandma, Tom and Joseph went swimming.

____

____

1 mark

11 Contract these words using an **apostrophe**.

should have ________________________

they will ________________________

he is ________________________

3 marks

12 Put a tick in each row to show whether each **explanation** is **true** or **false**.

Explanation	True	False
A **request** is to ask for something.		
An **indulgent** is an invitation.		
A **container** is used to put things in.		
A **bleak** morning is a **dismal** one.		

4 marks

13 Write the **basic form** of each verb.

laughs, laughed = to ______________________

walks, walking = to ______________________

flies, flew = to ______________________

grows, grew = to ______________________

1 mark

14 Rewrite the sentence below using **standard English**.

I've not gone none.

____

1 mark

15 What is the name of the **punctuation mark** below?

. . .

____ 1 mark

16 Underline the **modal verb** in the sentence below.

We should have stayed late at school tonight. 1 mark

17 Rewrite the sentence below, changing it to the **past tense**.

I eat all my dinner.

____ 1 mark

18 The sentence below has an **apostrophe** missing.

Rewrite the sentence and **explain** why the apostrophe is needed.

Annies mum worked at the school library.

____

____

____ 2 marks

Total ________/24 marks

Spelling

TEST 1

- There are 20 spellings and you will hear each spelling three times.
- Ask someone to read the instructions and sentences to you. These can be found on page 91.
- This test should take approximately 15 minutes but you will be allowed as much time as you need.

1 Jai hurt his ________________ playing tennis with his friends.

2 The ________________ forecast for today is mostly sunny and warm.

3 Our ________________ little kitten had scratched the carpet.

4 Mangoes, pineapples, kiwis and oranges are all types of ________________.

5 We are going on a ________________ hunt tomorrow.

6 The new clothes didn't fit so I need to ________________ them to the shop.

7 My ambition is to play football for my ________________.

8 The ________________ was extremely busy this afternoon.

9 At break time, we play with the outdoor ________________.

10 The ________________ water boils at is 100 degrees Celsius.

11 The doctor said the rash was highly ________________.

12 Sam won a ________________ he entered at school.

13 Jemima was an ________________ height for her age.

14 Our family eat ________________ for breakfast.

15 The actress was very ________________ on the stage.

16 The flowers in the bathroom are ________________.

17 The weather is very ________________ today.

18 There is a ________________ coming from the large window.

19 The building must be evacuated ________________ if the fire alarm rings.

20 The ________________ members all attended a meeting at the school last night.

Total ________/20 marks

TEST 2

Spelling

- There are 20 spellings and you will hear each spelling three times.
- Ask someone to read the instructions and sentences to you. These can be found on pages 91–92.
- This test should take approximately 15 minutes but you will be allowed as much time as you need.

1 The ______________________ of Majorca is close to Spain.

2 The chocolate cake was ______________________.

3 ______________________ can be mashed, chipped, boiled or baked.

4 A ______________________ should never be broken!

5 My ______________________ sport is cricket.

6 The ______________________ of the luggage is less than 15 kilograms.

7 ______________________ teachers have told me I have a talent for drawing.

8 The kittens are ______________________.

9 Playing football for the school team is a great ______________________.

10 All the ______________________ information should be underlined.

11 I am trying to ______________________ my mum to buy me some new boots.

12 My ______________________ feels full after lunch.

13 Mata listened to the story with ______________________.

14 The first time we met was a little ______________________.

15 Ned's ______________________ was the fastest.

16 The whole family is in ______________________ for my first pet.

17 Because of the roadworks, the traffic was ______________________.

18 Coffee and fizzy drinks usually contain ______________________.

19 The gymnasts ______________________ their weight across the bars.

20 We do not mind what type of ______________________ we stay in on holiday.

Total ________/20 marks

Contents

A Mad Tea Party

Alice has just drunk a magic potion, which has made her very small. It enables her to enter Wonderland, where nothing is ever quite as it seems!

The scene begins in the garden. There is a table under a tree in front of a house. The March Hare and the Hatter are having tea at the table. A dormouse sits between them.

Narrator: The March Hare and the Mad Hatter are having tea, all squashed at one end of a very large table.

Alice: Hello, mind if I join you all?

Hare, Hatter, Dormouse: No room! No room!

Alice: What are you talking about, there's plenty of room.

Alice sits down

Hare: Have some wine.

Alice: *(looking around the table)* I don't see any wine.

Hare: There isn't any.

Alice: *(in an angry voice)* Then it wasn't very civil of you to offer it!

Hare: It wasn't very civil of you to sit down without being invited.

Alice: I didn't know it was your table, it's laid for many more than three.

Hatter: *(looking up and down at Alice)* Your hair wants cutting.

Alice: You should learn not to make personal remarks. It's very rude!

Narrator: The characters continue arguing until the Hatter opens his eyes very wide and begins talking in riddles.

Hatter: Why is a raven like a writing desk?

Alice: I believe I can guess that.

Hare: Do you mean you think you can find the answer for it?

Alice: Exactly so.

Hare: Then you should say what you mean.

Alice: I do. At least, I mean what I say – that's the same thing you know.

Hatter: Not the same thing a bit! Why, you might just as well say that I like what I get is the same as I get what I like.

Dormouse: *(talking in a sleepy way)* I breathe when I sleep is the same thing as I sleep when I breathe!

Hatter: It is the same thing with you!

Narrator: The conversation drops, and the party sits silently for a minute.

(The Hatter takes his watch out of his pocket and looks at it uneasily; he shakes it and holds it to his ear. He then dips it in his cup of tea and looks at it again.)

James and the Giant Peach

The garden lay soft and silver in the moonlight. The grass was wet with dew and a million dewdrops were sparkling and twinkling like diamonds around his feet. And now suddenly, the whole place, the whole garden seemed to be alive with magic.

Almost without knowing what he was doing, as though drawn by some powerful magnet, James Henry Trotter started walking slowly toward the giant peach. He climbed over the fence that surrounded it, and stood directly beneath it, staring up at its great bulging sides. He put out a hand and touched it gently with the tip of one finger. It felt soft and warm and slightly furry, like the skin of a baby mouse. He moved a step closer and rubbed his cheek lightly against the soft skin. And then suddenly, while he was doing this, he happened to notice that right beside him and below him, close to the ground, there was a hole in the side of the peach.

It was quite a large hole, the sort of thing an animal about the size of a fox might have made.

James knelt down in front of it and poked his head and shoulders inside.

He crawled in.

He kept on crawling.

This isn't just a hole, he thought excitedly. It's a tunnel!

The tunnel was damp and murky, and all around him there was the curious bittersweet smell of fresh peach. The floor was soggy under his knees, the walls were wet and sticky, and peach juice was dripping from the ceiling. James opened his mouth and caught some of it on his tongue. It tasted delicious.

He was crawling uphill now, as though the tunnel were leading straight toward the very centre of the gigantic fruit. Every few seconds he paused and took a bite out of the wall. The peach flesh was sweet and juicy, and marvelously refreshing.

He crawled on for several more yards, and then suddenly – bang – the top of his head

bumped into something extremely hard blocking his way. He glanced up. In front of him there was a solid wall that seemed at first as though it were made of wood. He touched it with his fingers. It certainly felt like wood, except that it was very jagged and full of deep grooves.

'Good heavens!' he said. 'I know what this is! I've come to the stone in the middle of the peach!'

Then he noticed that there was a small door cut into the face of the peach stone. He gave a push. It swung open. He crawled through it, and before he had time to glance up and see where he was, he heard a voice saying, 'Look who's here!' And another one said, 'We've been waiting for you!'

James stopped and stared at the speakers, his face white with horror.

He started to stand up, but his knees were shaking so much he had to sit down again on the floor. He glanced behind him, thinking he could bolt back into the tunnel the way he had come, but the doorway had disappeared. There was now only a solid brown wall behind him.

James's large frightened eyes travelled slowly around the room.

The creatures, some sitting on chairs, others reclining on a sofa, were all watching him intently.

Creatures?

Or were they insects?

An insect is usually something rather small, is it not? A grasshopper, for example, is an insect.

So what would you call it if you saw a grasshopper as large as a dog? As large as a large dog. You could hardly call that an insect, could you?

There was an Old-Green-Grasshopper as large as a large dog sitting on a stool directly across the room from James now.

And next to the Old-Green-Grasshopper, there was an enormous Spider.

And next to the Spider, there was a giant Ladybird with nine black spots on her scarlet shell.

Each of these three was squatting upon a magnificent chair.

On a sofa near by, reclining comfortably in curled-up positions, there was a Centipede and an Earthworm.

On the floor over in the far corner, there was something thick and white that looked as though it might be a Silkworm. But it was sleeping soundly and nobody was paying any attention to it.

Every one of these 'creatures' was at least as big as James himself, and in the strange greenish light that shone down from somewhere in the ceiling, they were absolutely terrifying to behold.

'I'm hungry!' the Spider announced suddenly, staring hard at James.

'I'm famished!' the Old-Green-Grasshopper said.

'So am I!' the Ladybird cried.

The Centipede sat up a little straighter on the sofa. 'Everyone's famished!' he said. 'We need food!'

Four pairs of round black glassy eyes were all fixed upon James.

The Centipede made a wriggling movement with his body as though he were about to glide off the sofa – but he didn't.

There was a long pause – and a long silence.

The Spider (who happened to be a female spider) opened her mouth and ran a long black tongue delicately over her lips. 'Aren't you hungry?' she asked suddenly, leaning forward and addressing herself to James.

Poor James was backed up against the far wall, shivering with fright and much too terrified to answer.

'What's the matter with you?' the Old-Green-Grasshopper asked. 'You look positively ill!'

'He looks as though he's going to faint any second,' the Centipede said.

'Oh, my goodness, the poor thing!' the Ladybird cried. 'I do believe he thinks it's him that we are wanting to eat!' There was a roar of laughter from all sides.

'Oh dear, oh dear!' they said. 'What an awful thought!'

'You mustn't be frightened,' the Ladybird said kindly. 'We wouldn't dream of hurting you. You are one of us now, didn't you know that? You are one of the crew. We're all in the same boat.'

'We've been waiting for you all day long,' the Old-Green-Grasshopper said. 'We thought you were never going to turn up. I'm glad you made it.'

'So cheer up, my boy, cheer up!' the Centipede said. 'And meanwhile I wish you'd come over here and give me a hand with these boots. It takes me hours to get them all off by myself.'

The Isles of Scilly

Location

The Isles of Scilly are a group of islands located 45 km (28 miles) off the coast of the Cornish peninsula.

The Islands

The Isles of Scilly are made up of five inhabited and approximately 140 other islands. The largest of the islands is called St Mary's, with Tresco, St Martin's, St Agnes and Bryher being the others.

The main settlement on St Mary's is Hugh Town. This is a very small town with only a few shops, banks, restaurants, hotels and pubs. The population of Hugh Town is just over 1000.

Travel

Because the Isles of Scilly are so small, very few people own cars. Locals usually walk or ride bicycles. Public transport is a boat service between the islands.

To reach the Isles of Scilly from the mainland, visitors need to travel on either a small aeroplane (Skybus) or a passenger ferry (*The Scillonian*). These methods of transport are both an amazing experience for travellers to the islands and the table below lists the advantages and disadvantages of each.

	Skybus	***The Scillonian***
Advantages	Travel to and from three different mainland airports	Cafés serving hot and cold food on board
	Bird's-eye views of the islands	Often see dolphins, sea birds and basking sharks
	Quick	Seaside views
	Small and personal	Cheap
Disadvantages	Limited luggage allowed	Often rough seas and passengers may suffer motion sickness
	More expensive	Longer journey
	Flights can be cancelled because of high winds	

Weather

The Isles of Scilly have a unique climate, with the mildest and warmest temperatures in the United Kingdom. The average annual temperature is 11.8 °C (53.2 °F) in comparison to London, where it is 11.6 °C (52.9 °F).

The winters in Scilly are relatively warm and the islands very rarely get frosts or snow. They are however windy, due to the full force of the wind off the Atlantic Ocean.

Plants and Animals

The warm, humid climate of the Isles of Scilly allows a variety of rare plants and flowers to grow, which are not seen in other parts of the United Kingdom.

Some people describe the Scilly Islands as 'a large natural greenhouse'. Even in winter, there are hundreds of different plants in bloom.

Month	**Jan**	**Feb**	**Mar**	**Apr**	**May**	**Jun**	**Jul**	**Aug**	**Sept**	**Oct**	**Nov**	**Dec**
Average temp. °C (°F)	8.0 (46.4)	7.9 (46.2)	8.7 (47.8)	9.8 (49.6)	12.0 (53.5)	14.5 (58.1)	16.5 (61.7)	16.9 (62.5)	15.5 (59.8)	12.7 (54.8)	8.5 (50.6)	8.7 (47.6)

Because of the warm temperatures, many birds migrate to Scilly. Some birds stop off on their way across the Atlantic Ocean, while others stay on the islands. There are many different types of birds on the islands, ranging from common birds such as sparrows, thrushes and gulls, to more unusual birds like puffins, cuckoos and wheatears.

Tourism

The Isles of Scilly rely on tourism. The majority of shops, restaurants, bars and cafés earn money from visitors during the summer season. People travel to Scilly for many reasons. Quite often there are visitors who come on walking holidays, bird-watching trips and to relax on one of the quiet beaches.

Holidaymakers can choose accommodation suitable for their stay. The main islands have a choice of hotels, bed and breakfasts, self-catering cottages and camping facilities.

Activities

Many people chose to participate in water sports, including sailing, kayaking, boating and fishing.

The islands are famous for their clear waters, which also make snorkelling and diving popular activities.

'Gig racing' is the main sport on the Isles of Scilly. Gigs are traditional working boats, which six or seven people sit inside and row. Many islanders take part in gig practices and races throughout the season.

People watch the races and cheer on the gigs from passenger boats that follow the race or they sometimes watch the finish from the quay on St Mary's. The Isles of Scilly host an annual gig championship; thousands of rowers from as far away as Holland flock to the island to race at this event.

Historic Sites

The Isles of Scilly have many historic landmarks waiting to be explored, including castles, churches, lighthouses and ruins. There are art galleries and exhibitions, museums and often concerts to attend.

Probably the most impressive feature of the Isles of Scilly is its natural beauty; the swirling golden beaches, sand dunes, rocky coastlines and scenic views are well worth a visit.

http://www.collins.co.uk/floellabenjamin

About Floella

Floella Benjamin was born in the Caribbean on an island called Trinidad on 23rd September 1949. Her father decided to emigrate to England and she came to join him in 1960, when she was 11 years old. Floella's family started their life in Great Britain in Beckenham, South London. Initially she found it very different to her life in Trinidad and growing up in two cultures was challenging.

Floella's mother (Marmie) was a great inspiration to her. She gave her lots of love and encouraged her to do well at school.

Floella always dreamed of becoming a teacher, but instead ended up working in a bank and then later starred in stage musicals. While Floella enjoyed being on stage, she also wanted to try working in television so she auditioned for a variety of roles. Her TV debut was in 1974 in a drama called *Within these Walls* and her success in the show landed her many more roles in TV dramas.

Floella then took on a new role as a presenter of *Playschool*, a 1970s children's show. She loved this job and presented the programme for 12 years.

As well as acting and presenting, Floella also loves to sing. She began singing with her dad who had a jazz band.

She often sang with her dad's band and also with large classical orchestras.

In 1987 Floella started her own television production company.

Next page ▶▶▶

http://www.collins.co.uk/floellabenjamin

Floella has also starred in several pantomimes, worked on numerous radio programmes, narrated audio books and has done voiceovers for a range of adverts and commercials.

Since 1983, Floella has written over 25 children's books.

One book Floella is particularly proud of is one that has been made into a film. *Coming to England* was written in 1995 and is based on her own life. In the book Floella talks about what it's like to be different, to move countries and change cultures and her feelings of rejection. The drama *Coming to England* won a Royal Television Society Award in 2004.

Floella is a keen runner who has completed ten London Marathons for the children's charity Barnardo's. This achievement is even more special as until she turned 50 years old, Floella had never even ran at all!

Trekking the Great Wall of China in 2004 in aid of NCH Action for Children is another one of Floella's achievements. She started this in the Gobi Desert and finished 400 km later where the Great Wall meets the Yellow Sea.

Today Floella's passion is for inspiring and helping children and young people. She is a patron and supporter of many charities, including Action for Children and the Sickle Cell Society. In 2008, Floella was inducted into the National Society for the Prevention of Cruelty to Children (NSPCC) Hall of Fame.

Floella supports and empathises with children and young people across the world. Through her charity work, her writing and personal attitudes, she strives to help young people find their identity and to understand where they come from. She hopes all children have a sense of belonging and learn to become proud of themselves.

Street Child

'Joseph,' Jim asked the bent man one day out in the yard. 'How long have you been here?'

'Been here?' Joseph swung his head round and peered up at Jim. 'Seems like I was born here. Don't know nothing else, son. And I don't know all of this place neither.' He leaned against Jim so he could swing his head up to look at the long, high building with its rows of barred windows. 'I've not been in the room where the women go, though long ago I must have been in the baby-room, I suppose, with my ma. I've been in the infirmary wards. But there's all kinds of little twisty corridors and attics and places I've never been in Jim, and I don't want to, neither. It's the whole world, this place is.' He spread out his hands. 'Whole world.'

'It ain't, Joseph,' Jim told him. 'There's no shops here, and no carriages. And no trees.' He closed his eyes, forcing himself to try and remember what it was like outside. 'And there's no river. There's a great big river outside here.'

'Is there now?' said Joseph. 'I should like to see that river. Though to tell you the truth, Jim, I don't know what a river is. Tell you something.' He put his arm over Jim's shoulder to draw his ear closer to his own mouth. 'I don't want to die in here. If someone will let me know what day I'm going to die, I'll be grateful. I'll climb over that wall first.' He dropped his head down again and stared at his boots, whistling softly. 'Yes. That's what I'll do.'

Tip spluttered and nudged Jim, but Jim was looking up at the high walls that surrounded the workhouse, and at the bleak sky above it.

'How long have I been here, Tip?' he asked.

'How should I know?' Tip hugged his arms round himself. 'Keep moving, Jim. It's cold.'

It was impossible to tell one day from the next. They were all the same. School, sack-making, bed. The only thing that changed was the sky. Jim had seen the grey of snow clouds turning into the soft rain clouds of spring. He'd felt summer scorching him in his heavy, itchy clothes. And now the sky was steely grey again. The pump had long beards of ice on its handle.

'I've been here a year,' Jim said. It was then that the little secret promise that had nestled inside him began to flutter into life like a wild thing.

'I've got to skip off,' he let the mad thought rise up in him.

'If I don't, I'll be like Joseph. One day I won't remember whether I was born here or not. I won't know anywhere but here.'

During lessons that day, the old schoolmaster's voice droned on in the dim schoolroom. The boys coughed and shuffled in their benches, hunching themselves against the cold. Jim's wild thoughts drummed inside him, so loud that he imagined everyone would hear them. He leaned over to Tip and whispered in his ear, 'Tip, I'm going to run away today. Come with me?' Tip sheered round, and put his hand to his mouth. Mr Barrack sprang down from his chair, his eyes alight with anger and joy.

'You spoke!' he said to Jim, triumphant. 'It was you.'

Tip closed his eyes and held out his hand, but Jim stood up. He didn't mind. He didn't mind anything any more. The teacher hauled him off his stool and swung his rope round. It hummed as it sliced through the air.

'I don't mind,' Jim tried to explain, but this made Mr Barrack angrier than ever. At last he had caught Jim out, and he was beating him now for every time he had tried and failed. He pulled a greasy handkerchief out of his pocket and wound it round Jim's head, tying it tight under his chin.

'Just in case you feels like hollering,' he said. All the other boys stared in front of them. The rope stung Jim again and again and the beating inside him was like a wild bird now, throbbing in his limbs and in his stomach, in his chest and in his head, so wild and loud that he felt it would lift him up and carry him away.

When the schoolmaster had finished with him he flung him like a bundle of rags across the desk. Jim lay in a shimmer of pain and thrumming wings. He wanted to sleep. The bell rang and the boys shuffled out. Jim felt Tip's hand on his shoulder. He flinched away.

'That's what they do to the boys who skip off, Jim,' Tip whispered. 'They thrash 'em like that every day until they are good.'

Jim felt the wild thing fluttering again. 'Only if they catch them.'

'They always catch 'em. Bobbies catch 'em and bring 'em in, and they get thrashed and thrashed.'

Jim struggled to sit up. The stinging rolled down his body. 'Won't you come with me?'

'I daresn't. Honest, I daresn't. Don't go, Jim.'

Jim looked up at the great archways of the schoolroom. He knew the words off by heart. God is good. God is holy. God is just. God is love.

'I've got to,' he said. 'And I'm going tonight, Tip.'

Jim knew that he would have to make his break before old Marion did her rounds for the night. He had no idea how he was going to do it. At suppertime he stuffed his cheese in his pocket, and Tip passed his own share along to him.

At the end of the meal, Mr Sissons stood up on his dais. All the shuffling and whispering stopped. He moved his body slowly round, which was his way of fixing his eyes on everyone, freezing them like statues.

'I'm looking for some big boys,' he said. 'To help with the carpet-beaters.' He waited in silence, but nobody moved.

'Just as I would expect. A rush to help, when there is sickness in the wards,' A cold sigh seemed to ripple through the room. Mr Sissons laughed into it in his dry, hissing way. 'It might be cholera, my dears. That's what I hear. I've two thousand mouths to feed here, and someone has to earn the money, cholera or not. Somebody has to buy the medicines. Somebody has to pay for the burials.' He moved his body round in its slow, watchful circle again. 'Plenty of big strong boys here, eating every crumb I give them and never a word of thanks.' He stepped down from his dais and walked along the rows, cuffing boys on the back of their heads as he passed them. 'I want you all up in the women's wards straight after supper, and you don't come down again till all the carpets are done.'

'What's carpets?' asked Jim.

'Dunno,' Tip whispered, 'they come from the rich houses, and the women here beat 'em, and then they send them home.'

'I'm going with them,' Jim said suddenly, standing up as soon as the older boys did.

'A daft boy, you are,' said Tip. 'He asked for big boys.'

'You coming or not?' Jim darted off after the big boys and Tip ran after him.

Macavity's a Mystery Cat: he's called the Hidden Paw –
For he's the master criminal who can defy the Law.
He's the bafflement of Scotland Yard, the Flying Squad's despair:
For when they reach the scene of crime – Macavity's not there!

Macavity, Macavity, there's no one like Macavity,
He's broken every human law, he breaks the law of gravity.
His powers of levitation would make a fakir stare,
And when you reach the scene of crime – Macavity's not there!
You may seek him in the basement, you may look up in the air –
But I tell you once and once again, Macavity's not there!

Macavity's a ginger cat, he's very tall and thin;
You would know him if you saw him, for his eyes are sunken in.
His brow is deeply lined with thought, his head is highly domed;
His coat is dusty from neglect, his whiskers are uncombed.
He sways his head from side to side, with movements like a snake;
And when you think he's half asleep, he's always wide awake.

Macavity, Macavity, there's no one like Macavity,
For he's a fiend in feline shape, a monster of depravity.
You may meet him in a by-street, you may see him in the square –
But when a crime's discovered, then Macavity's not there!

He's outwardly respectable. (They say he cheats at cards.)
And his footprints are not found in any file of Scotland Yard's.
And when the larder's looted, or the jewel-case is rifled,
Or when the milk is missing, or another Peke's been stifled,
Or the greenhouse glass is broken, and the trellis past repair
Ay, there's the wonder of the thing! Macavity's not there!

And when the Foreign Office find a Treaty's gone astray,
Or the Admiralty lose some plans and drawings by the way,
There may be a scrap of paper in the hall or on the stair –
But it's useless to investigate – Macavity's not there!
And when the loss has been disclosed, the Secret Service say:
'It must have been Macavity!' – but he's a mile away.
You'll be sure to find him resting, or a-licking of his thumbs,
Or engaged in doing complicated long division sums.

Macavity, Macavity, there's no one like Macavity,
There never was a Cat of such deceitfulness and suavity.
He always has an alibi, and one or two to spare:
At whatever time the deed took place – MACAVITY WASN'T THERE!
And they say that all the Cats whose wicked deeds are widely known,
(I might mention Mungojerrie, I might mention Griddlebone)
Are nothing more than agents for the Cat who all the time,
Just controls their operations: the Napoleon of Crime!

Reading Tests: Answers

Test 1: *A Mad Tea Party*

1. Accept either in the garden or in Wonderland. **(1 mark)**
2. She has just drunk a magic potion that makes her small. **(1 mark)**
3. *...there's plenty of room.* **(1 mark)**
4. the Hare **(1 mark)**
5. polite **(1 mark)**
6. '...all squashed at one end of a very large table.' **(1 mark)**
7. Any one of the following points:
   - Alice thinks personal comments are rude.
   - The Hatter's comments are said very abruptly and to the point.
   - The Hatter's comments are unprovoked. **(1 mark)**
8. The Mad Hatter begins talking in riddles. **(1 mark)**
9. Any one of the following: lowers; they become quiet; ends. **(1 mark)**
10. Any two from: confident; friendly; determined; argumentative; sensible.
    (2 marks: award 1 mark for each description)
11. Any two from: a narrator; new line for each speaker; stage directions; use of direct language; set as a scene.
    (2 marks: award 1 mark for each correct answer)
12. **(a)** uneasily **(1 mark)**
    (b) Answers will vary. Examples: mad, rude, unpredictable, unusual. **(1 mark)**
13. Any three from: it is a tea party which is very unusual; the characters are acting in strange ways; they are saying mad things; they are sitting squashed up; they are talking in riddles; they are offering wine which isn't there.
    (3 marks: award 1 mark for each description)
14. Any one from: Alice sits down; looking around the table; in an angry voice; looking up and down at Alice; talking in a sleepy way. **(1 mark)**
15. Alice's **(1 mark)**
16. puzzle **(1 mark)**
17. the Hare, Mad Hatter and Dormouse **(1 mark)**
18. Any two from: the Narrator sets the scene; moves the story forward; explains what's happening; introduces characters.
    (2 marks: award 1 mark for each correct answer)
19. *An Unusual Tea Party* **(1 mark)**

Test 2: *James and the Giant Peach*

1. a simile **(1 mark)**
2. Any one of the following: terrified; scared; worried; unsure; frightened. **(1 mark)**
3. closely **(1 mark)**
4. 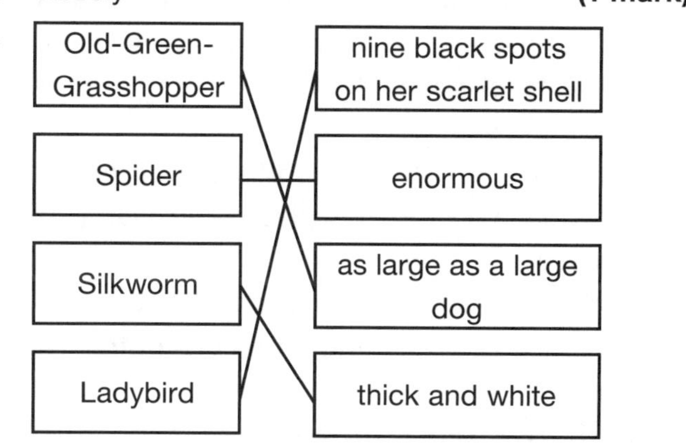

   (1 mark)
5. Any one from: charge back; race back; go very quickly. **(1 mark)**
6. The Ladybird thinks James is thinking that the creatures are going to eat him. **(1 mark)**
7. You have joined us now, you are just the same as us and we will treat you as our friend / family. You are no different from us – we are all here together. **(1 mark)**
8.

	True	False
A fence surrounded the peach.	✓	
The tunnel was dry and cold.		✓
James bumped his head on the stone in the middle of the peach.	✓	
James couldn't stand up because the peach wasn't big enough.		✓
The creatures were at least the same size as James.	✓	

(1 mark: all five correct for 1 mark)

9. Author is creating suspense. What is going to happen next? Telling us there were four creatures staring, the word 'fixed' creates the image – they were all looking at James, nothing else. The author is helping readers understand the anxiety James must be feeling.
   (2 marks: 1 mark for each point made)

10. Use of personification – 'the garden lay', 'the whole garden seemed to be alive with magic'. Use of the word 'moonlight' – creating atmosphere, implying that something exciting, unusual, magical is about to happen. Use of similes – visual imagery such as 'sparkling and twinkling like diamonds'.
(2 marks: 1 mark for each point made)
11. The Centipede has many feet so has a lot of boots to take on and off. **(1 mark)**
12. head and shoulders
(1 mark: both correct for 1 mark)
13. Answers will vary. Examples: damp; wet. **(1 mark)**
14. excited **(1 mark)**
15. Old-Green-Grasshopper and Ladybird
(1 mark: both correct for 1 mark)

Test 3: *The Isles of Scilly*

1. St Mary's **(1 mark)**
2. people live there **(1 mark)**
3. a small aeroplane (Skybus)
a passenger ferry (*The Scillonian*)
(1 mark: both correct for 1 mark)
4. A day visitor with no luggage would be best taking the aeroplane (Skybus) because it's quicker and they would have longer to spend in the Isles of Scilly. They have no luggage so can travel by aeroplane. They will get a good view of all the islands even though they will not get time to see them all on foot. They will be able to choose one of three airports to travel from on the mainland.
(2 marks, 1 mark for each correct point, up to 2 marks)
A family staying for a week who want to see dolphins swimming would be best taking the passenger ferry because they may be able to see dolphins (and basking sharks and rare birds) from the ferry. They are staying a whole week so have plenty of time to spend travelling to Scilly by boat. The family might want to buy food or drinks on board and, as the boat is cheaper, it will cost them less.
(2 marks, 1 mark for each correct point, up to 2 marks)
5. It is a warm, humid, tropical place. The weather is predictable and hundreds of flowers and plants are always in bloom. **(1 mark)**
6. There are lots of them, they are not unusual or rare. **(1 mark)**
7.

Language used	Explanation of the effect of the language
. . . **flock** to the island to race.	It makes the reader think rowers group like a flock of sheep or birds, and that there are lots of them.
. . . **waiting** to be explored . . .	This phrase gives readers an image of the buildings waiting for people to visit them – really the buildings are just standing there – the author adds interest.
. . . **swirling golden** beaches . . .	Conjures up a picture of the wind gently blowing the sand, which swirls into dunes. 'Golden' makes the reader think of a sunny, warm day, making the beach inviting.

(3 marks)

8.

Statement	Reason
Very few people own cars.	Because the Isles of Scilly are so small.
The islands are very windy.	Due to the full force of the wind off the Atlantic Ocean.
A variety of rare plants and flowers grow.	There is a warm, humid climate.
Birds migrate to the Isles of Scilly.	Because of the warm temperatures.

(4 marks: award 1 mark for each correct answer)

9. Any of the following answers are acceptable but they must be consistently either 'yes' or 'no'.
Yes:
- They are small.
- Not many cars – good to get away from lots of traffic.
- Enjoy boats, walking or cycling.
- Lots of wildlife – birds, fish, plants to see.
- Lots of tourism.
- Relaxing; peaceful; quiet.
- Lovely beaches, sand dunes, rocky coastlines.
- Varied accommodation.
- Water sports available.
- Want to watch the gig races.
- Would like to explore castles, churches, lighthouses.
- Would like to visit art galleries, exhibitions, museums, attend concerts.

- Beautiful islands.
- Amazing views.
- Warm climate.

No:

- Difficult to get to – you can't drive.
- Very quiet – not many people live there.
- The main town does not have many large facilities.
- Not keen on water, boats or water sports.
- Would prefer somewhere not as warm.
- No interest in birds or wildlife.
- Not much to do.

(1 mark: all three correct for 1 mark)

10. Participating: a participant would sit inside the gig; row; take part in races.
Watching: a spectator would watch the race from a boat; or the quay on St Mary's; cheer on the gigs.
(2 marks: 1 mark for one participating point, 1 mark for one watching point)

11. Any one suitable question, e.g. What is the weather like in the winter? Where do you go to work? How often do you visit the mainland? **(1 mark)**

12. Any two of the following:

- maps
- charts
- headings / subheadings
- present tense
- facts are used
- clear language. **(2 marks)**

13. Any three from: walking; bird watching; relaxing on beaches; water sports; visiting historic sites; enjoying the wildlife / plants.
(2 marks: 1 mark for two correct answers, 2 marks for three correct answers)

14. Hugh Town **(1 mark)**

15. Any two from: walk; ride bicycles; by boat.
(2 marks: award 1 mark for each correct answer)

16. a travel leaflet **(1 mark)**

Test 4: *About Floella*

1. biography **(1 mark)**

2. Accept either in the Caribbean or Trinidad. **(1 mark)**

3. Any one from: very difficult; hard work; a struggle. **(1 mark)**

4. Mum: Floella's mother (Marmie) was a great inspiration to her; gave her lots of love; encouraged her.
Dad: She began singing with her dad; she often sang with her dad's jazz band.
(2 marks: 1 mark for point made for mum, 1 mark for one point made for dad)

5. **(a)** *Coming to England* **(1 mark)**
(b) Floella came to England – it is based on her own life and she knows what it's like to be different, to come to a new country and to have to adjust to living in a different culture. **(1 mark)**

6. *Within these Walls* **(1 mark)**

7.

Event	Date
Born	1949
Came to England	1960
Made her TV debut	1974
Started her own television production company	1987
Wrote *Coming to England*	1995
Trekked the Great Wall of China	2004

(2 marks: 1 mark if four or five answers are correct, 2 marks if all six answers are correct)

8. chronological **(1 mark)**

9. Barnardo's **(1 mark)**

10. Any two from: factual; pictures; true statements.
(2 marks: award 1 mark for each answer)

11. Completed ten London Marathons for Barnardo's; trekked the Great Wall of China for NCH Action for Children and the Sickle Cell Society; worked for, and raised money for, the NSPCC.
(3 marks: 1 mark for each correct answer)

12.

	Fact	Opinion
She worked in a bank.	✓	
She loved her job on *Playschool*.		✓
She presented *Playschool* for 12 years.	✓	
She has completed ten London Marathons.	✓	
She is a keen runner.		✓

(1 mark: all correct for 1 mark)

13. Because Floella moved to a new culture when she was a child she is aware of the difficulties children face in this situation; she wants to help children in a similar position; to help them find their identity; and for them to understand where they came from.
(3 marks: 1 mark for each point made, up to 3 marks)

Test 5: *Street Child*

1. A workhouse **(1 mark)**
2. He doesn't know life anywhere else. **(1 mark)**
3. **(a)** Any one from: so that the children can't escape; so that the children can't see out; so no one can see in. **(1 mark)**
   (b) Any one from: what life is like outside the workhouse; if he would ever see life outside the workhouse again. **(1 mark)**
4. He could tell the time of year by the weather; the colour of the sky; the light, and dark mornings and nights; he is aware of the change in the seasons.
   (2 marks: 1 mark for each point, up to 2 marks)
5. He can't stay here forever or for another year; he needs to escape so he will try to run away.
   (2 marks: 1 mark for each point, up to 2 marks)
6. pulled **(1 mark)**
7. Any one from: he knew he had plans to run away; he wasn't going to be there much longer, he had more important things to think about; this would be his last beating so he felt positive and thought he could deal with anything. **(1 mark)**
8. He is afraid they'll get caught and get thrashed.
   (2 marks: 1 mark for worried they'll get caught, 1 mark for fear of being thrashed)
9.

Language used	Explanation of the effect of the language
Jim's wild thoughts **drummed** inside him . . .	The thoughts were pounding away – beating / niggling / constantly there inside him – they were like a drum beat – loud / strong / constant.
. . . the **beating** inside him was like a wild bird now . . .	His thoughts were getting stronger – like a wild bird's wings – they were spreading out – he was thinking more about it – he couldn't contain his thoughts any more – they needed to escape and fly like a bird – be released.
. . . a shimmer of pain and **thrumming** wings.	A sudden burst of pain – 'his wings' had been damaged due to the beating he had just endured – he was hurt.

(3 marks: 1 mark for each correct answer)

10. **Yes:** for thinking he can escape the workhouse; for running away; for thinking he can go with the big boys – he'll get caught and then get another terrible beating; he has no real plans – where will he sleep? How will he eat?
    No: for having dreams; for trying to find a way out of the workhouse.
    (2 marks: 1 mark for each point, up to 2 marks)
11. Look out for each other: 'Don't go, Jim'.
    Care for each other: 'Tip passed his own share along to him'.
    Trust each other: Jim told Tip his plans – 'Tip, I'm going to run away today. Come with me?'
    Strong friendship: they suffer the workhouse together.
    (2 marks: award 1 mark for each correct point)
12. He doesn't remember / know life anywhere else
    (1 mark)

13.

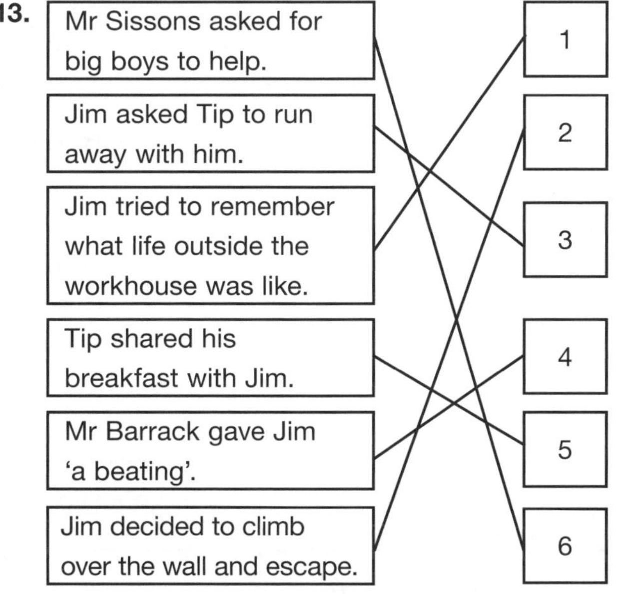

(3 marks: 1 mark for two correct answers, 2 marks for four correct answers, 3 marks for six correct answers)

14. joy, triumphant **(1 mark: both correct for 1 mark)**

15. Any one from: in case he feels like hollering; in case he shouts or makes a noise. **(1 mark)**

Test 6: *Macavity*

1. a cat **(1 mark)**

2. He's not there. **(1 mark)**

3. appearance **(1 mark)**

4. Any one from: uncared for; left alone; not looked after. **(1 mark)**

5. Whenever a crime has been committed Macavity is always involved but he hides away to keep out of trouble so his whereabouts are a mystery to those looking for him. **(1 mark)**

6. Macavity's absence is one of the main points of the poem. Repetition of this phrase emphasises this point. **(1 mark)**

7. Any sensible idea that fits in with Macavity's character. Examples:

Crime	What do you think Macavity did?
The larder's looted.	Stole some tuna. Took some biscuits – left a mess behind.
The jewel-case is rifled.	Jumped on the case and knocked it over, made a mess.
The milk is missing.	Drank it all.
The greenhouse glass is broken.	Jumped onto a plant pot and it fell over and broke a window.

(4 marks: award 1 mark for each correct answer)

8. **(a)** He always has an explanation for where he was at the time of the crime – sometimes he has more than one explanation. **(1 mark)**

(b) Any one from: he has planned the crimes; they are organised. **(1 mark)**

9. Any sensible questions, e.g. Where is your favourite hiding place?; How did you break the greenhouse?; Why do you behave in this way?

(2 marks: award 1 mark for each question)

10. narrative **(1 mark)**

11. The poet uses rhyme for effect – it is regular throughout each verse. Rhyming couplets are used. The poem has even rhyme, which helps the poem to be read / performed easily.

(2 marks: 1 mark for each point, up to 2 marks)

12. By the end of the poem, the poet is fed up of telling us that Macavity is not there . . . we know by now that he wasn't there. **(1 mark)**

13.

	Fact	Opinion
It is fun watching Macavity cause trouble.		✓
Macavity's a ginger cat.	✓	
His coat is dusty.	✓	
Macavity has no-one to care for him.		✓

(1 mark: all correct for 1 mark)

14. 'movements like a snake' **(1 mark)**

15. looted **(1 mark)**

16. Any two from: Macavity acts as the boss, the ringleader; he controls the crimes the other cats do; he's the best; the one in charge.

(2 marks: award 1 mark for each correct answer)

17. Any three suitable answers from:

- Macavity runs away from trouble.
- Macavity is a naughty cat – he breaks all the rules.
- He is a street cat – he always looks neglected.
- He is outwardly respectable – he tricks people.
- He controls the other cats – he's in charge.
- Macavity is sly.
- It is always Macavity who has committed or organised the crime.

Grammar and Punctuation Tests: Answers

Test 1

1. an elephant; a superhero
   (1 mark: both correct for 1 mark)
2. discover **(1 mark)**
3. We (was / were) going to play outside on our bikes.
   I (done / did) well in my maths test this morning.
   (2 marks: 1 mark for each correct part of the question)
4. 'Wait for me!' shouted Orla.
   'Ouch! My hand hurts.'
   Freddie whispered, 'Are you scared?'
   (3 marks: 1 mark for each correct answer)
5. Olivia and Sandip were building sandcastles at the beach. **(1 mark)**
6. The authors wrote letters to the newspaper. **(1 mark)**
7. Nottingham is located in the East Midlands. The river that runs through Nottingham is called the River Trent.
   (2 marks: 1 mark for correct use of capital letters and 1 mark for correct use of full stops)
8. Since it was very cold outside, Sam decided to fasten his coat. **(1 mark)**
9. 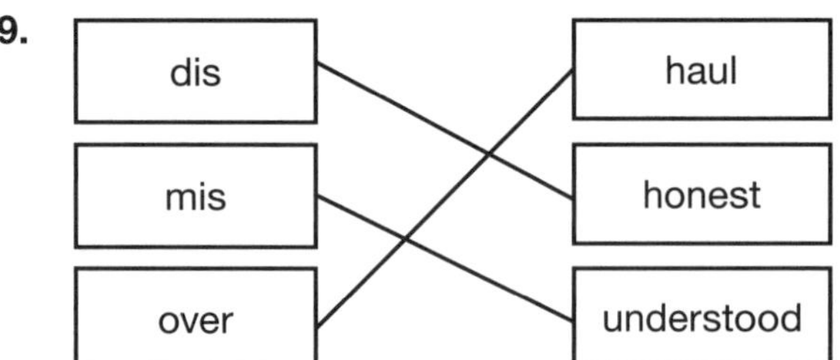

   (1 mark: all correct for 1 mark)
10. Emma's cat had five kittens last night.
    The boys' changing rooms were locked.
    Jamie's lunchbox was left overnight in his classroom.
    (3 marks: 1 mark for each correct answer)
11. William Shakespeare – a famous author – wrote the play 'Macbeth'. **(1 mark)**
12. Although she is younger than me, my sister is much taller. **(1 mark)**
13. Any suitable adverbial of time, e.g. First; Then; Next.
    (2 marks: 1 mark for each correct adverbial)
14. You've been learning about materials in science, haven't you? **(1 mark)**
15. While the band was playing, I was dancing with my friends. **(1 mark: both correct for 1 mark)**
16. therefore **(1 mark)**
17. Feeling confident, the pianist played to a room full of people. **(1 mark)**
18. colon **(1 mark)**
19. pronoun **(1 mark)**
20. I will wash the car today **(1 mark)**

Test 2

1. 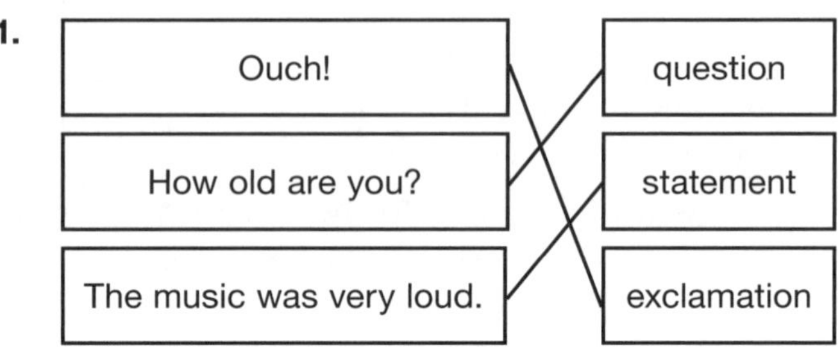

   (1 mark: all correct for 1 mark)
2. Dad slowly walked to the shops. **(1 mark)**
3. Circle round the word he. **(1 mark)**
4. Later that evening, we said goodbye and began our journey home. **(1 mark)**
5. because **(1 mark)**
6. Mangoes, kiwis, apples, pineapples and strawberries are all types of fruit.
   (1 mark: all correct for 1 mark)
7. We laughed at each other's funny jokes. **(1 mark)**
8. I am allowed to watch television while eating my dinner. **(1 mark)**
9. Answers may vary. Examples:

	Synonym	Antonym
sad	unhappy	happy
rich	wealthy	poor
big	large	small
old	ancient	new

   (3 marks: 1 mark for each correct line)
10. The water park was **an** amazing sight. It had **a** wave machine and **a** slide and **the** big pool had lots of toys.
    (1 mark: all correct for 1 mark)
11. The skateboard park, located behind the playground, is to be used by children who are over eight. **Or** The skateboard park (located behind the playground) is to be used by children who are over eight. **(1 mark)**
12. On your marks . . . ready . . . steady . . . go! **(1 mark)**
13. The historians visited the ancient ruins. **(1 mark)**

14.

Noun	Verb
simplification	to simplify
suffocation	to suffocate
magnification	to magnify

(3 marks)

15. The girl walked up the stairs. **(1 mark)**
16. Any correct addition of a subordinate clause will be awarded marks, e.g. Despite the cold, rainy morning, the farmer went into his field; The farmer, who owned several acres of land, went into his field. **(2 marks: 1 mark for correct addition of a subordinate clause and 1 mark for correct use of punctuation)**
17. Any correct sentence that uses a modal verb will be awarded marks, e.g. might, should, will, must. **(1 mark)**
18. The old lady who was shouting at her neighbour was feeling angry. **(1 mark)**
19. 'You're eating all your dinner, aren't you?' **(1 mark)**

Test 3

1. Remember to walk down the stairs quietly and carefully shut the door. **(2 marks: 1 mark for each correct answer)**
2. Wednesday; Lydia; We; Spanish **(1 mark: all correct for 1 mark)**
3. anti **(1 mark)**
4. Put; Mix **(1 mark: both correct for 1 mark)**
5. He **(1 mark)**
6. At the supermarket I bought milk, eggs, cheese and apples. **(1 mark)**
7. Any one from: can't; won't; didn't; don't; couldn't; wouldn't. **(1 mark)**
8. When I danced in the show I smiled all the time. [to dance → danced; to smile → smiled] **(2 marks: 1 mark for each correct word)**
9. 'What time does the train leave the station?' Mary asked the guard.
   '10:03,' the guard answered. **(2 marks: 1 mark for each correct line)**
10. The bird is going to look for food. **(1 mark)**
11. 'Where are you going' the bus driver asked. [✓ under going'] **(1 mark)**
12. The books **and** DVDs need returning to the library on Monday **or** Tuesday **but** remember it is closed each day for lunch. **(1 mark: all correct for 1 mark)**
13. Feeling a little anxious, the children walked into the hall. **(1 mark)**
14. The weather forecast is good for every day next week. **(1 mark: both correct for 1 mark)**
15. While I was waiting to see the doctor, I read a book. **(2 marks: 1 mark for correct fronted adverbial, 1 mark for correct use of comma)**
16. The dog chewed the ball. **(1 mark)**
17. If the bus were late again, the children would be cross. **(1 mark)**
18. The cat – fast asleep on the rug – was keeping warm by the fire. **(1 mark)**
19. With his sports kit on, Tom was ready for the game. **(1 mark)**
20. There was a crash ... a loud bang ... then darkness! **(1 mark)**
21. The vase was knocked over by the cat. **(1 mark)**

Test 4

1. The creepy shadows darted quickly past the wooden fence, along the hidden path and into the gloomy garage. **(1 mark: all correct for 1 mark)**
2. because **(1 mark)**
3. did; have **(2 marks: 1 mark for each correct answer)**
4. The dog's tail was wagging.
   The boys' coats were on the floor.
   The women's changing rooms were busy. **(3 marks: 1 mark for each correct sentence)**
5. 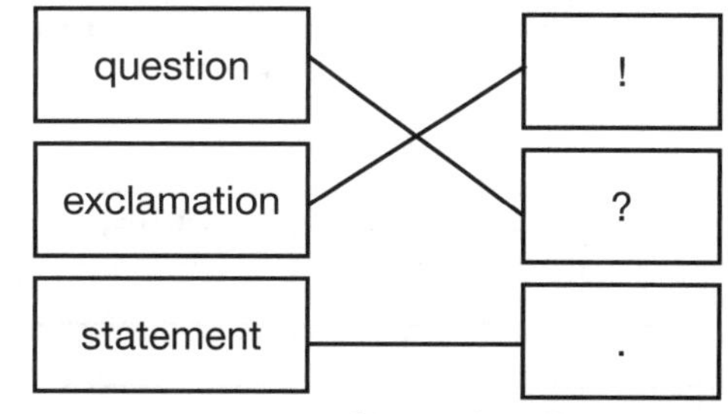

(1 mark: all correct for 1 mark)

6. After the singing, Kenzie blew out the candles. **(1 mark)**
7. Any correct answer, e.g.
   The busy skateboard park.
   The skiing holiday. **(2 marks: 1 mark for each correct sentence)**
8. I want to eat James – this sounds like the person actually wants to eat James.
   I want to eat, James – the person is telling James they are hungry; they want something to eat. **(1 mark)**
9. ellipsis **(1 mark)**
10. Grace bought a bouncy ball; she played with it in the garden. **(1 mark)**

11.

Sentence	Proper noun	Common noun
My birthday is in **January**.	✓	
The **bees** landed on the flower.		✓
The car towed the **caravan**.		✓
His brother's name is **Andrew**.	✓	

(1 mark: all correct for 1 mark)

12. should; ought (to) **(1 mark)**
13. noun = 'suffocation' **(1 mark)**
14. Rosie climbed the stairs. — active; The stairs were climbed by Rosie. — passive

(1 mark: both correct for 1 mark)

15. The girls were tired (but) it was too early to go to bed. **(1 mark)**
16. 'You were expecting this letter today, (weren't you?)' **(1 mark)**
17. The ice cream van that is in the car park is very popular today. **(1 mark)**
18. She walked along the canal path.
The dog hid under the table.
I was lost in the maze. **(3 marks)**

Test 5

1.

	Singular	Plural
children		✓
dog	✓	
women		✓
bananas		✓

(1 mark: all correct for 1 mark)

2. The delicious juicy apple fell from the large tree.
(1 mark: all correct for 1 mark)
3. The little boy watched TV and ate a sandwich. **(1 mark)**
4. (An) alligator is (a) large reptile.
(1 mark: both correct for 1 mark)
5. Rosie's homework was too hard for **her** and **she** felt sad that **she** couldn't do it.
(3 marks: 1 mark for each correct pronoun)
6. He will go to school tomorrow. **(1 mark)**
7. Any suitable answer, e.g. isn't it?; don't you? **(2 marks)**
8. Mrs Jones (the Year 5 teacher) played the piano in school today. **(1 mark)**
9. Any suitable subordinate clauses, e.g.
Although we didn't have much time, I went to the park, **which is close to my friend's house**.
(1 mark)
10. Hanging upside down, the bat made loud noises as the night sky grew darker. **(1 mark)**
11. The dog ate the leftover food. **(1 mark)**
12. Any suitable words, e.g.

Word	Synonym	Antonym
rich	wealthy	poor
angry	cross	calm

(2 marks: 1 mark for each word if both the synonym and antonym are correct)

13. The tree (sway / (sways)) gently in the breeze.
The goats ((have) / has) two horns each.
The children ((are) / is) too noisy!
(1 mark: three correct for 1 mark)
14. One item (scissors, card, glue) written on each line.
(1 mark: 1 mark if all bullet points are correct)
15. Any three from:
dis / re / un cover
re elect
dis allow
(3 marks: 1 mark for each correct prefix)
16. I like drawing **and** painting **but / although** my sister prefers writing.
(1 mark: both correct for 1 mark)
17. After tea, we went to the cinema.
(2 marks: award 1 mark for correct adverbial and 1 mark for correct punctuation)
18. Jump over, Ben – implies someone is telling Ben to jump over something.
Jump over Ben – implies someone is telling others to jump over Ben. **(1 mark)**
19. Any one from: 'Be quiet!' Mrs Shepherd told her class before they went into assembly;
Mrs Shepherd told her class, 'Be quiet' before they went into assembly. **(1 mark)**

Test 6

1. Isla listened carefully (so) she understood the instructions. **(1 mark)**
2. The milkman delivers fresh produce on a Monday, Tuesday, Thursday and Saturday.
(1 mark: both correct for 1 mark)

3. Noah and **I** went for a walk in the woods.
'You pushed **me** into the mud.' **(1 mark)**
4. The cat slept (in) the warm, cosy basket. **(1 mark)**
5. Ella played the piano; Alex played the flute. **(1 mark)**
6.

Adverbial	Time (when)	Place (where)
frequently	✓	
regularly	✓	
often	✓	
into the bin		✓

(1 mark: all correct for 1 mark)
7. Brighton, which is a seaside town, is located on the south coast of England. **(1 mark)**
8. 'It's my turn next,' I told my older brother. **(1 mark)**
9. Any sentence that uses brackets correctly, e.g. The cat hid (in a large box) under the bed; The newsreader was walking away from her car (a Ford Fiesta) when she heard a loud noise. **(1 mark)**
10. The first sentence tells readers that just Tom and Joseph went swimming.
The second sentence tells us that Grandma, Tom and Joseph went swimming. **(1 mark)**
11. should've; they'll; he's
(3 marks: 1 mark for each correct word)
12.

Explanation	True	False
A **request** is to ask for something.	✓	
An **indulgent** is an invitation.		✓
A **container** is used to put things in.	✓	
A **bleak** morning is a **dismal** one.	✓	

(4 marks: 1 mark for each correct answer)
13. to laugh; to walk; to fly; to grow
(1 mark: all correct for 1 mark)
14. I have not got any. **(1 mark)**
15. Ellipsis **(1 mark)**
16. We should have stayed late at school tonight.
(1 mark)
17. I ate all my dinner. **(1 mark)**
18. Annie's mum worked at the school library.
The apostrophe is used for possession to show that the mum belongs to Annie.
(2 marks: 1 mark for correct use of apostrophe and 1 mark for correct explanation)

Spelling Tests: Answers

Test 1

These are the correct spellings:

1. shoulder
2. weather
3. naughty
4. fruit
5. treasure
6. return
7. country
8. supermarket
9. equipment
10. temperature
11. infectious
12. competition
13. average
14. cereal
15. confident
16. artificial
17. changeable
18. draught
19. immediately
20. committee

Test 2

These are the correct spellings:

1. island
2. delicious
3. potatoes
4. promise
5. favourite
6. weight
7. various
8. adorable
9. opportunity
10. relevant
11. persuade
12. stomach
13. curiosity
14. awkward
15. vehicle
16. mourning
17. stationary
18. caffeine
19. transferred
20. accommodation

Spelling Test Administration

The instructions below are for the spelling tests.
Read the following instruction out to your child:
I am going to read 20 sentences to you. Each sentence has a word missing. Listen carefully to the missing word and fill this in the answer space, making sure that you spell it correctly.
I will read the word, then the word within a sentence, then repeat the word a third time.
You should now read the spellings three times, as given below. Leave at least a 12-second gap between spellings. At the end, read all the sentences again, giving your child the chance to make any changes they wish to their answers.

Test 1

Spelling 1: The word is **shoulder**.
Jai hurt his **shoulder** playing tennis with his friends.
The word is **shoulder**.

Spelling 2: The word is **weather**.
The **weather** forecast for today is mostly sunny and warm.
The word is **weather**.

Spelling 3: The word is **naughty**.
Our **naughty** little kitten had scratched the carpet.
The word is **naughty**.

Spelling 4: The word is **fruit**.
Mangoes, pineapples, kiwis and oranges are all types of **fruit**.
The word is **fruit**.

Spelling 5: The word is **treasure**.
We are going on a **treasure** hunt tomorrow.
The word is **treasure**.

Spelling 6: The word is **return**.
The new clothes didn't fit so I need to **return** them to the shop.
The word is **return**.

Spelling 7: The word is **country**.
My ambition is to play football for my **country**.
The word is **country.**

Spelling 8: The word is **supermarket**.
The **supermarket** was extremely busy this afternoon.
The word is **supermarket**.

Spelling 9: The word is **equipment**.
At break time, we play with the outdoor **equipment**.
The word is **equipment**.

Spelling 10: The word is **temperature**.
The **temperature** water boils at is 100 degrees Celsius.
The word is **temperature**.

Spelling 11: The word is **infectious**.
The doctor said the rash was highly **infectious**.
The word is **infectious**.

Spelling 12: The word is **competition**.
Sam won a **competition** he entered at school.
The word is **competition**.

Spelling 13: The word is **average**.
Jemima was an **average** height for her age.
The word is **average**.

Spelling 14: The word is **cereal**.
Our family eat **cereal** for breakfast.
The word is **cereal**.

Spelling 15: The word is **confident**.
The actress was very **confident** on the stage.
The word is **confident**.

Spelling 16: The word is **artificial**.
The flowers in the bathroom are **artificial**.
The word is **artificial**.

Spelling 17: The word is **changeable**.
The weather is very **changeable** today.
The word is **changeable**.

Spelling 18: The word is **draught**.
There is a **draught** coming from the large window.
The word is **draught**.

Spelling 19: The word is **immediately**.
The building must be evacuated **immediately** if the fire alarm rings.
The word is **immediately**.

Spelling 20: The word is **committee**.
The **committee** members all attended a meeting at the school last night.
The word is **committee**.

Test 2

Spelling 1: The word is **island**.
The **island** of Majorca is close to Spain.
The word is **island**.

Spelling 2: The word is **delicious**.
The chocolate cake was **delicious**.
The word is **delicious**.

Spelling 3: The word is **potatoes**.
Potatoes can be mashed, chipped, boiled or baked.
The word is **potatoes**.

Spelling 4: The word is **promise**.
A **promise** should never be broken!
The word is **promise**.

Spelling 5: The word is **favourite**.
My **favourite** sport is cricket.
The word is **favourite**.

Spelling 6: The word is **weight**.
The **weight** of the luggage is less than 15 kilograms.
The word is **weight**.

Spelling 7: The word is **various**.
Various teachers have told me I have a talent for drawing.
The word is **various**.

Spelling 8: The word is **adorable**.
The kittens are **adorable**.
The word is **adorable**.

Spelling 9: The word is **opportunity**.
Playing football for the school team is a great **opportunity**.
The word is **opportunity**.

Spelling 10: The word is **relevant**.
All the **relevant** information should be underlined.
The word is **relevant**.

Spelling 11: The word is **persuade**.
I am trying to **persuade** my mum to buy me some new boots.
The word is **persuade**.

Spelling 12: The word is **stomach**.
My **stomach** feels full after lunch.
The word is **stomach**.

Spelling 13: The word is **curiosity**.
Mata listened to the story with **curiosity**.
The word is **curiosity**.

Spelling 14: The word is **awkward**.
The first time we met was a little **awkward**.
The word is **awkward**.

Spelling 15: The word is **vehicle**.
Ned's **vehicle** was the fastest.
The word is **vehicle**.

Spelling 16: The word is **mourning**.
The whole family is in **mourning** for my first pet.
The word is **mourning**.

Spelling 17: The word is **stationary**.
Because of the roadworks, the traffic was **stationary**.
The word is **stationary**.

Spelling 18: The word is **caffeine**.
Coffee and fizzy drinks usually contain **caffeine**.
The word is **caffeine**.

Spelling 19: The word is **transferred**.
The gymnasts **transferred** their weight across the bars.
The word is **transferred**.

Spelling 20: The word is **accommodation**.
We do not mind what type of **accommodation** we stay in on holiday.
The word is **accommodation**.

Progress Report

Fill in your total marks for each completed test.
Colour the stars to show how you feel after completing each test.
☆ = needs practice ☆☆ = nearly there ☆☆☆ = got it!

Reading

Test	Marks	How do you feel?
Test 1	/ 25	☆ ☆ ☆
Test 2	/ 17	☆ ☆ ☆
Test 3	/ 28	☆ ☆ ☆
Test 4	/ 21	☆ ☆ ☆
Test 5	/ 24	☆ ☆ ☆
Test 6	/ 26	☆ ☆ ☆

Grammar and Punctuation

Test	Marks	How do you feel?
Test 1	/ 27	☆ ☆ ☆
Test 2	/ 24	☆ ☆ ☆
Test 3	/ 25	☆ ☆ ☆
Test 4	/ 24	☆ ☆ ☆
Test 5	/ 26	☆ ☆ ☆
Test 6	/ 24	☆ ☆ ☆

Spelling

Test	Marks	How do you feel?
Test 1	/ 20	☆ ☆ ☆
Test 2	/ 20	☆ ☆ ☆

KS2 Success

Age 7-11

Science

Tests

Thomas Finch

Contents

Biology

- You have 25 minutes to complete this test.

TEST 1

1 Plant structure

a Label the diagram of the plant.

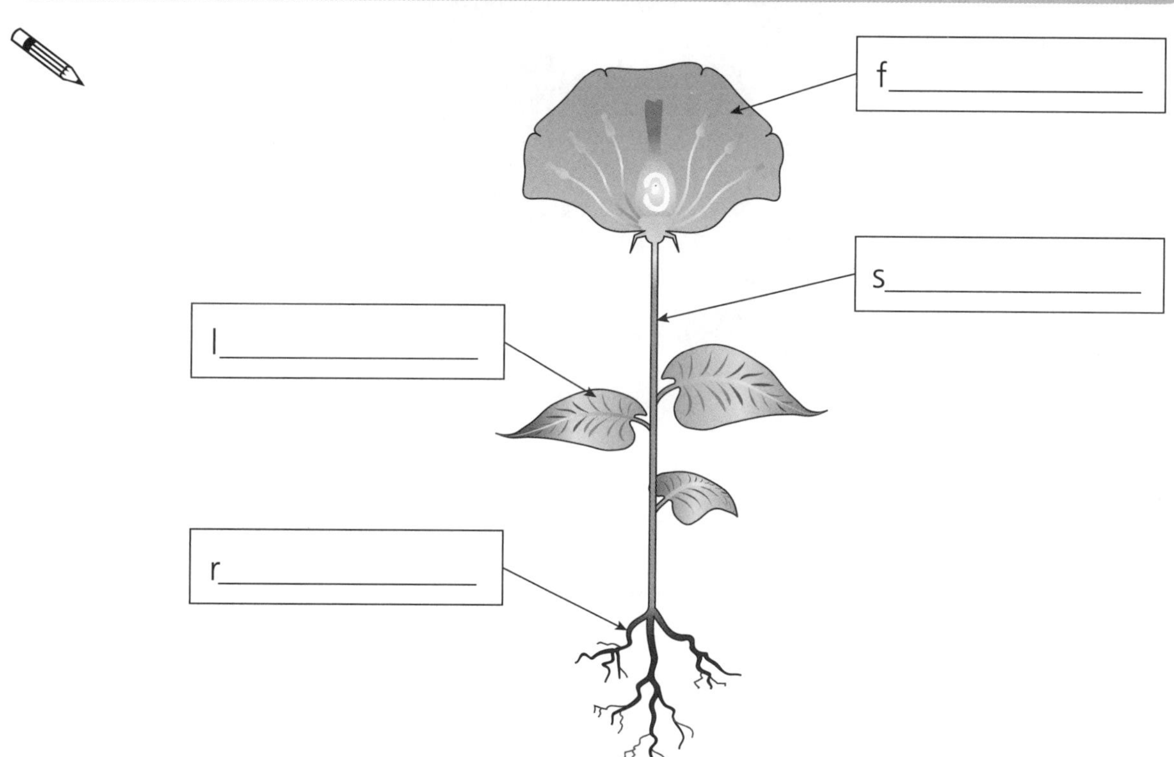

1 mark

b Draw a line to match each part of the flower with its function.

Stamen	The female part where seeds are made
Nectaries	Where nectar is made
Petal	The male part where pollen is made
Carpel	Attracts insects to the flower
Sepals	Protect the flower while it is a bud

3 marks

2 Teeth

a Bob is enjoying his tea. He knows that different teeth have different purposes when eating.

Explain the job of each type of tooth listed below.

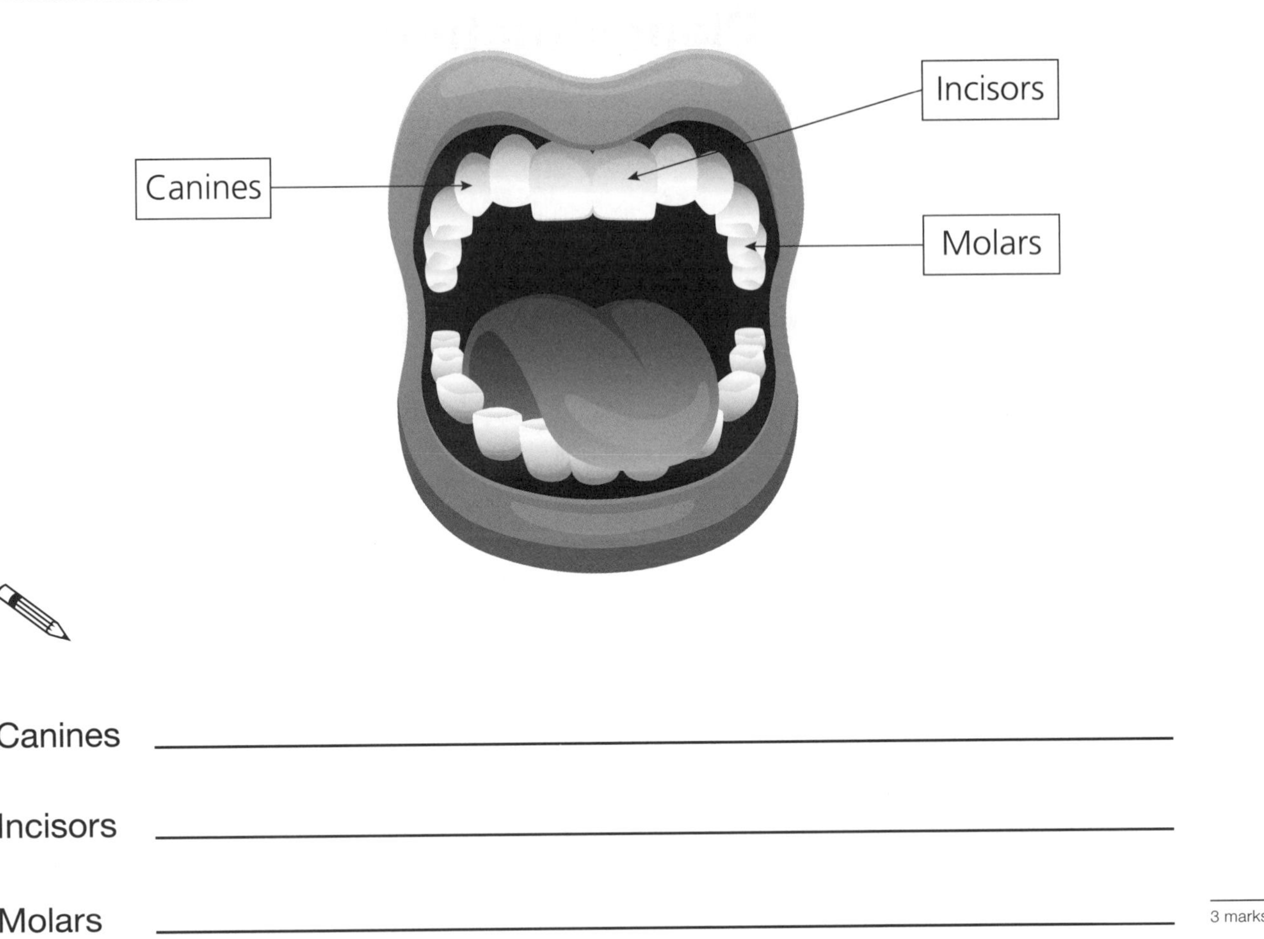

Canines ________________________________

Incisors ________________________________

Molars ________________________________

3 marks

b Bob knows that he needs to keep his teeth healthy.

Suggest **TWO** things Bob must do to keep his teeth healthy.

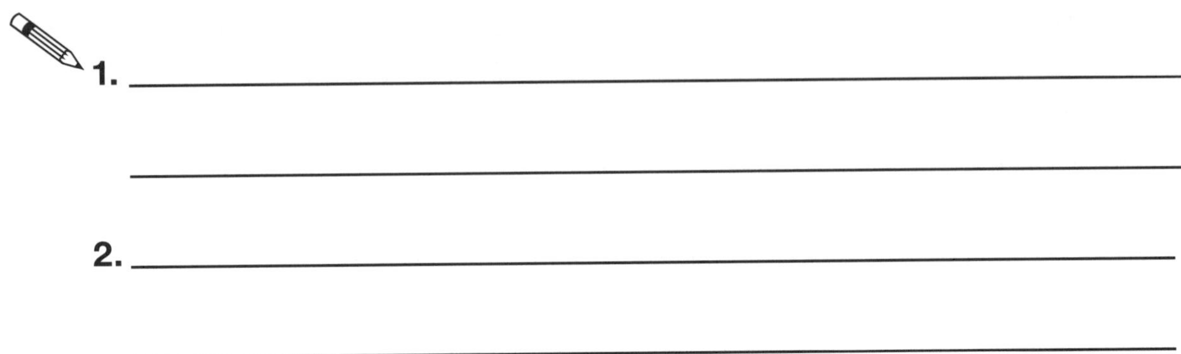

1. ________________________________

____

2. ________________________________

____

2 marks

3 Animal adaptations

a Animals have adapted to suit their environments over many years.

Draw a line to match each animal with its environment.

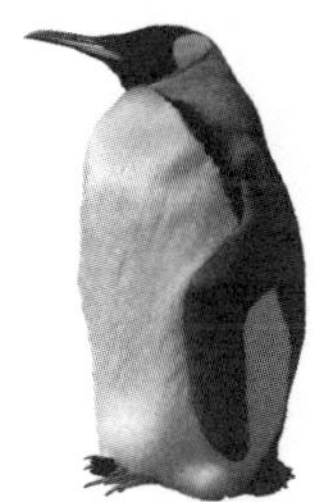

1 mark

b Choose **ONE** of the animals above and give **ONE** way it has adapted to live in its environment.

Animal: ______________________________

How has it adapted? ______________________________

____

1 mark

4 Genetics

a This picture shows some dogs.

Give **THREE** ways in which the dogs show variation.

1. ______________________________

2. ______________________________

3. ______________________________

3 marks

b Children are not usually identical to their parents.

Name **TWO** ways that show that children are not identical to their parents.

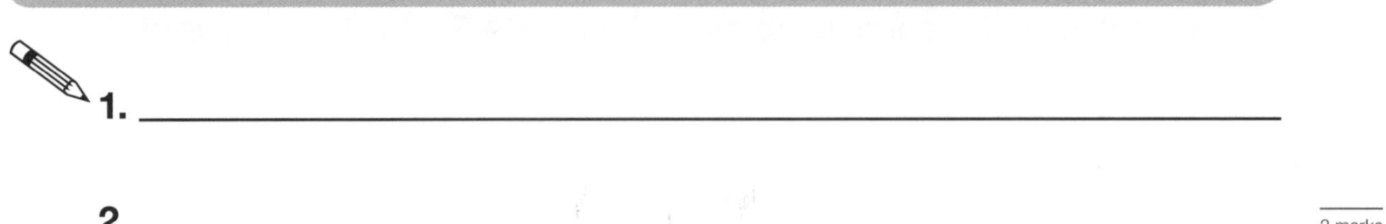

1. ______________________________

2. ______________________________

2 marks

5 The heart

a Identify the different parts of the circulatory system indicated in the diagram below.

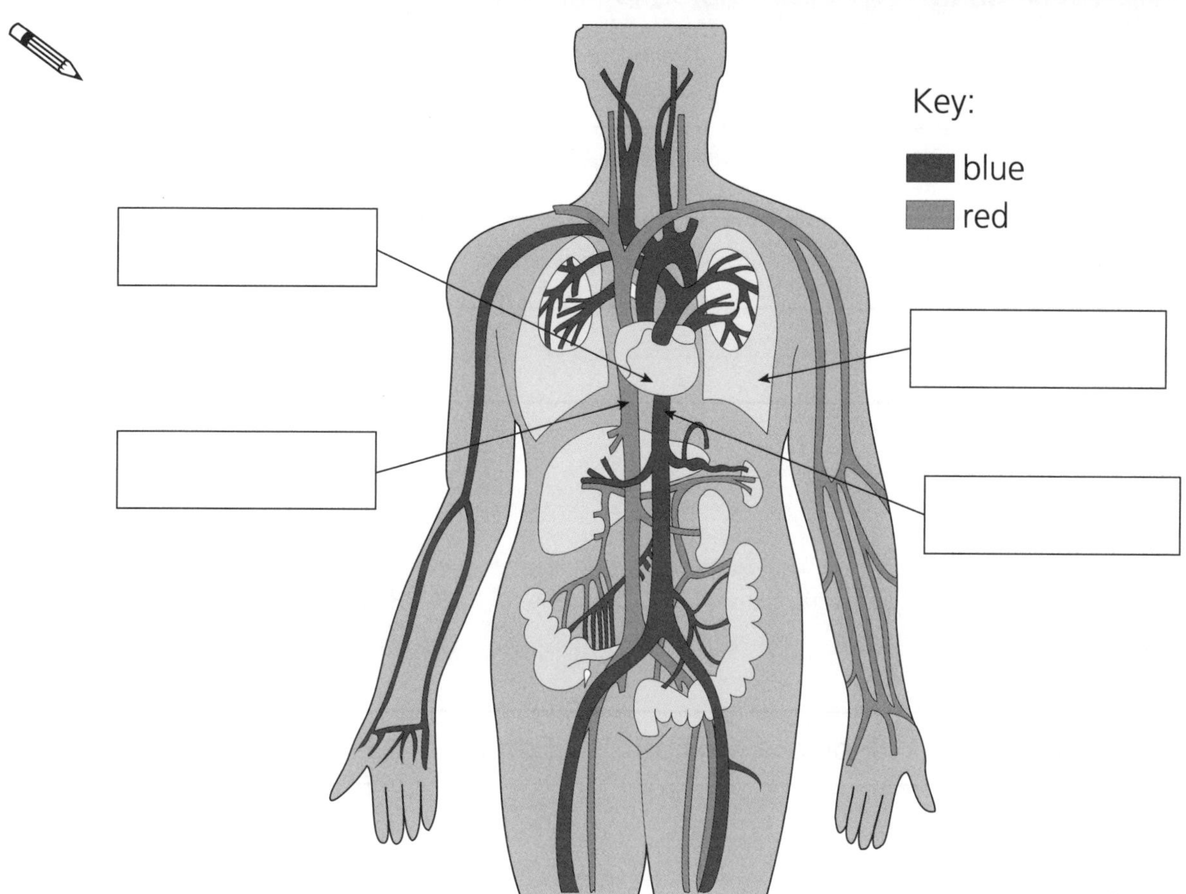

3 marks

b Draw a line to match each part of the circulatory system with its function.

Arteries	Tubes that carry blood to the heart
Veins	Carries oxygen around the body
Blood	Pumps blood around the body
Heart	Tubes that carry blood away from the heart

2 marks

6 The human skeleton

a All humans have a bony skeleton.

Give **TWO** reasons why a skeleton is important.

1. ____

____

2. ____

____

2 marks

b Use the words in the box below to identify the different parts of the skeleton.

Cranium (skull)	Fibula	Ribs	Femur	Humerus	Tibia

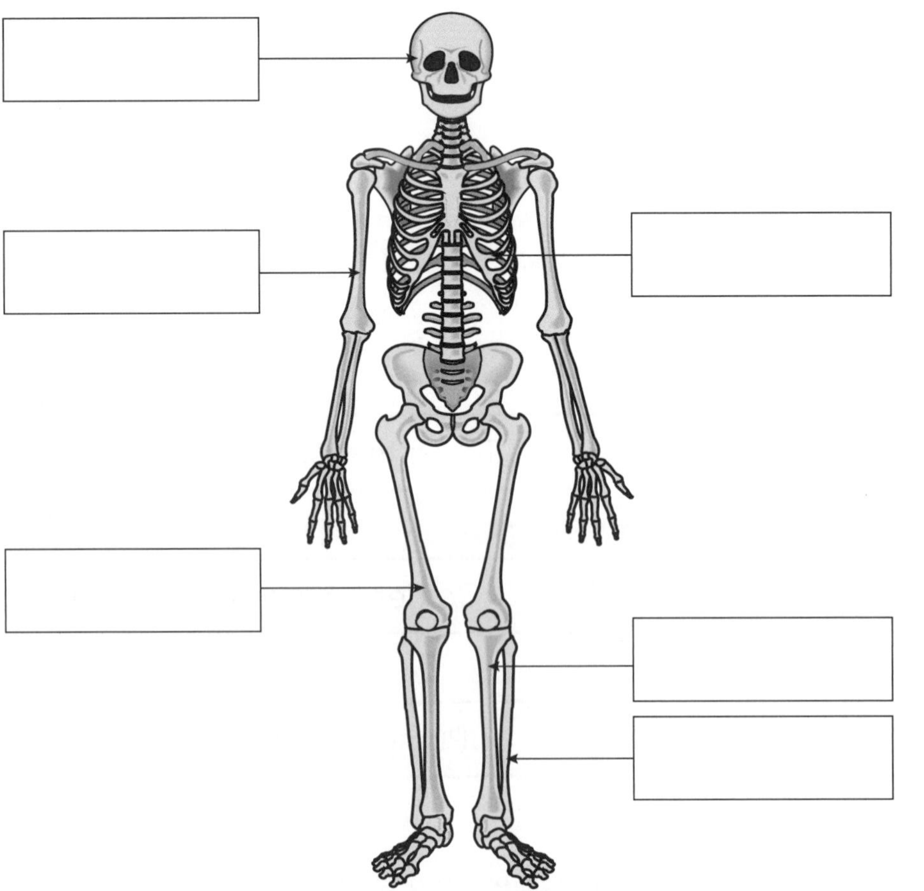

2 marks

Total ________/25 marks

Biology

- You have 25 minutes to complete this test.

TEST 2

1 Plant growth

a Plants need different things to live, including water.

Explain why plants need water.

____

____ 1 mark

b Where does water enter the plant?

____

____ 1 mark

c How does water get to the leaves and flowers?

____

____ 1 mark

2 Healthy lifestyles

a These people answered a survey on their lifestyles.

	Smoker	Healthy diet	Exercise
Ron	Yes	No	No
Kuba	No	Yes	No
Paul	Yes	Yes	Yes
Glen	No	Yes	Yes

Which person has the healthiest lifestyle?

____ 1 mark

b Why is smoking bad for smokers' lungs?

____

____ 1 mark

c Explain why it is important to have a healthy lifestyle.

____

____ 1 mark

3 Playing football

a Some children are playing football. Before they start playing, the children measure their pulses.

What does the pulse measure?

____ 1 mark

b After the children have been playing for 15 minutes, they measure their pulses again.

What will they notice?

____ 1 mark

c The children need to increase their breathing while doing exercise.

Why is it important to do this?

____

____

____ 2 marks

4 Food chains

a Mia is working on her vegetable patch and has observed the following:

Slugs eat the cabbages.
Thrushes eat the slugs.

Show this on a food chain.

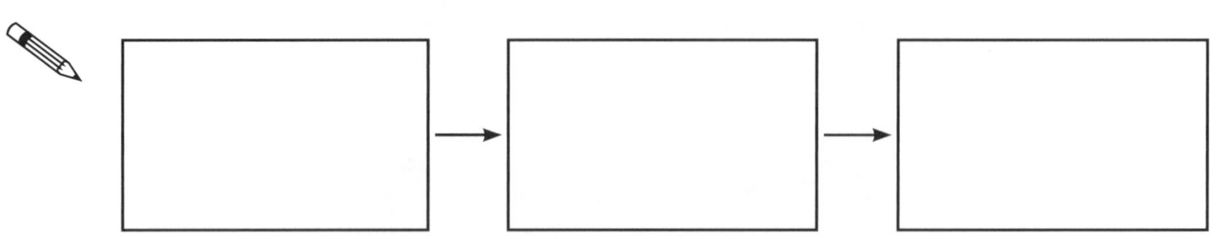

1 mark

b Match the follow things with the correct description.

Cabbage

Prey

Thrush

Producer

Slug

Predator

3 marks

5 Digestion

a The digestive system is important to make sure we get the nutrients we need.

Write a number from 1 to 4 in each box to order these stages in the digestive system from start to finish.

☐ Excess water is absorbed back into the body in the large intestine.

☐ Digested food is absorbed into the bloodstream through the small intestine.

☐ Any undigested food passes out when we go to the toilet.

☐ Food is eaten, then digested in the mouth, stomach and small intestine.

2 marks

b Tick the name of the tube in which chewed food passes from the mouth to the stomach.

Small intestine ☐

Large intestine ☐

Oesophagus ☐

Rectum ☐

1 mark

6 Animals and their environments

a Animals have changed over millions of years.

Write down **ONE** feature of a polar bear that helps it in its environment.

______________________________ 1 mark

b How does this help the polar bear in its environment?

____

______________________________ 1 mark

c Giraffes have developed long necks over time.

Give **TWO** advantages to the giraffe of having a long neck.

1. ______________________________

____

____

2. ______________________________

____

______________________________ 2 marks

d Over time the polar bears' environment has started to melt away.

How is this affecting polar bears?

____

____

______________________________ 2 marks

Total ______ / 23 marks

Biology

TEST 3

- You have 25 minutes to complete this test.

1 Plants

a Alice is planning an investigation on plants. She has set up three different plants around the room in different places.

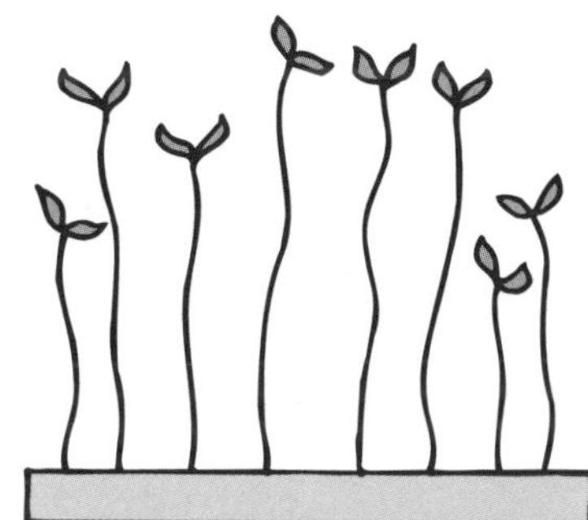

1. Next to the window

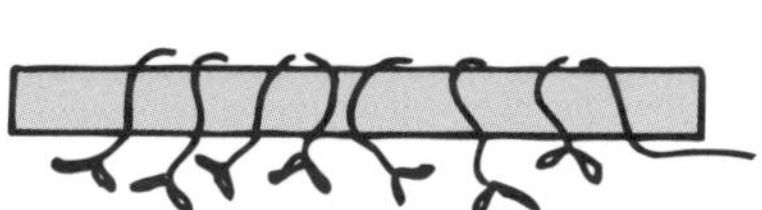

2. Inside a cupboard

3. Inside a plastic bag in the middle of the room

Alice waters each plant once a day.

What is plant 2 missing in order to grow well?

____

____ 1 mark

b Name all **THREE** things that plants need to survive.

1. ____

2. ____

3. ____ 3 marks

2 Muscles

a Humans have muscles throughout their bodies.
A group of children are getting ready for a PE lesson.

Why do the children need to warm up and stretch their muscles before exercising?

____

____

1 mark

b The biceps muscle in the picture below is currently relaxed.

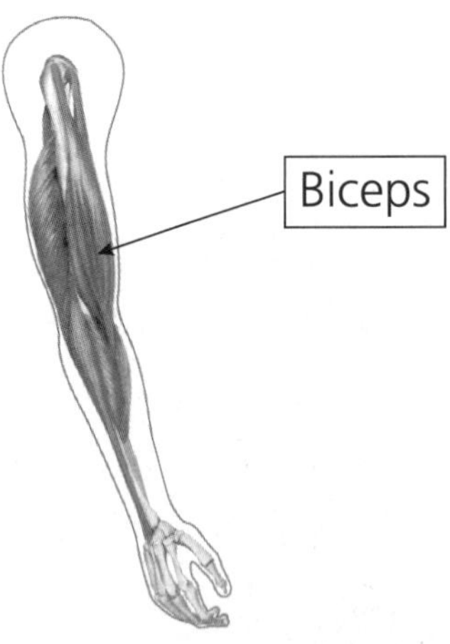

Explain what happens when this muscle is contracted.

____

____

____

2 marks

3 Life cycles of animals

a Explain the difference between vertebrate and invertebrate animals.

1 mark

b Sort the following animals into the correct columns in the table.

Mammals	Birds	Spiders	Insects

Vertebrate	Invertebrate

2 marks

c Draw the life cycle of a penguin.

1 mark

4 Life cycles of plants

a Plants need to spread their seeds in order to make new plants.

Order the events in this process from 1 to 3.

☐ Seeds are scattered by an animal or the wind.

☐ Pollen is blown by the wind or carried by insects from one plant to another.

☐ Pollen reaches the carpel of the new flower and travels to the ovary. It then fertilises ovules to make seed.

2 marks

b Identify the name of each event in the process.

(i) Seeds are scattered by an animal or the wind.

____

1 mark

(ii) Pollen is blown by the wind or carried by insects from one plant to another.

____

1 mark

(iii) Pollen reaches the carpel of the new flower and travels to the ovary. It then fertilises ovules to make seed.

____

1 mark

c Describe the function of these parts of the plant.

Petal ____________________

Stamen ____________________

Carpel ____________________

3 marks

d Explain what germination is.

____

____

____

1 mark

e Tick the part of the plant that takes in water and nutrients.

Leaves ☐

Roots ☐

Stem ☐

Flower ☐

1 mark

5 Human life

a A baby is breathing in his cot.

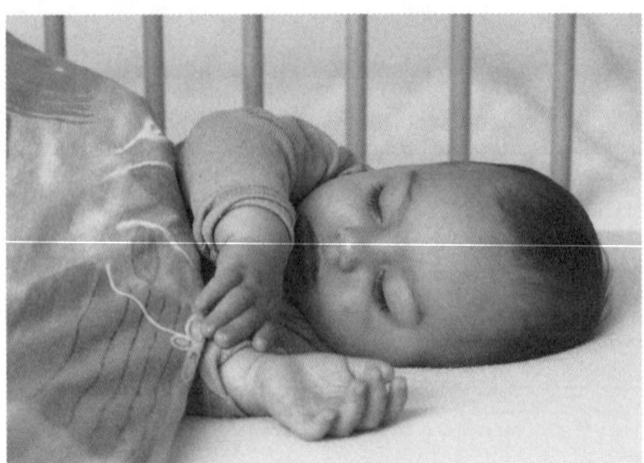

Tick the statements which show the baby is a living thing.

☐	Grows	☐	Drinks milk
☐	Eats	☐	Urinates
☐	Wears a hat	☐	Cries
☐	Sits on the floor	☐	Has a bath

2 marks

b Complete the following life cycle of a human.

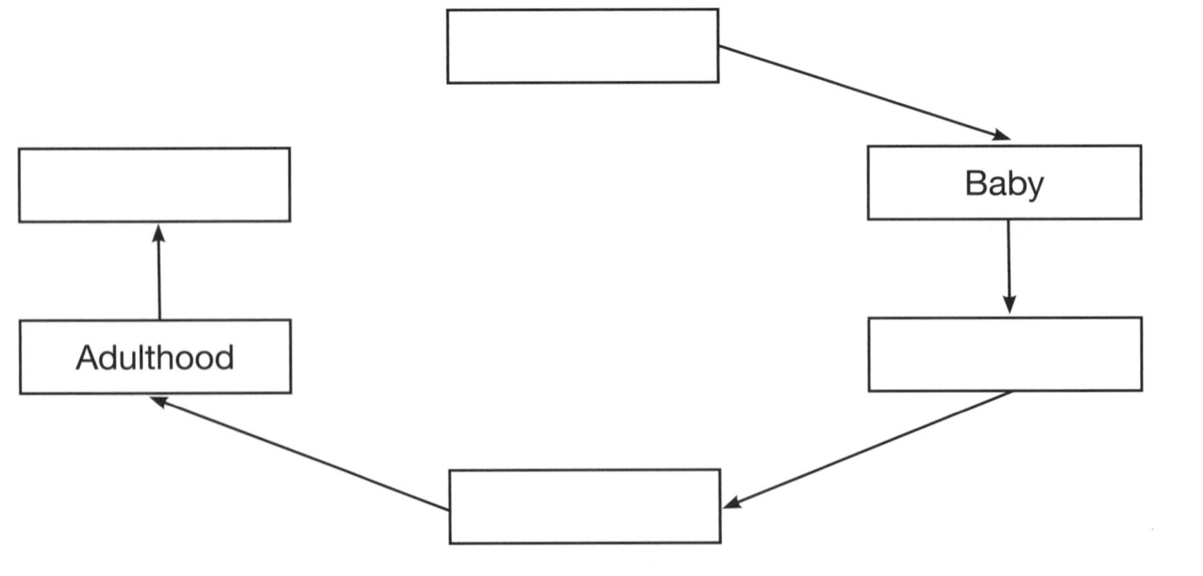

2 marks

Total ________/25 marks

Chemistry

TEST 1

- You have 25 minutes to complete this test.

1 Fire

a Hashim is making a fire. He lights a match to set fire to the wood.

What **THREE** things are formed when the wood burns?

1. ______________________

2. ______________________

3. ______________________

1 mark

b What kind of material change is this?

Irreversible	Reversible	Evaporating	Melting

____

1 mark

2 Fossils

a This is a fossil of an animal from millions of years ago.

What material are fossils made of?

______________________________ 1 mark

b Write a number from 1 to 5 in each box to order the events that took place to make this fossil.

☐	Rock pushes the sediment down and water washes away the bones, leaving a space in the rock.
☐	Fossils are uncovered millions of years later.
☐	Sediment falls onto the skeleton of the animal.
☐	The animal dies and drops to the river bed.
☐	Water carries rock into the area where the animal was, creating a fossil.

3 marks

c Scientists can gather information from fossils.

Write down **TWO** things that scientists can learn by examining a fossil.

1. ______________________________

2. ______________________________ 2 marks

3 Ice cubes

a Sid is enjoying a cold drink. As he drinks it, the ice cubes begin to melt.

Label each arrow with the name of the change that is happening.

One has been done for you.

Melting	Evaporating	Freezing	Boiling	Condensing

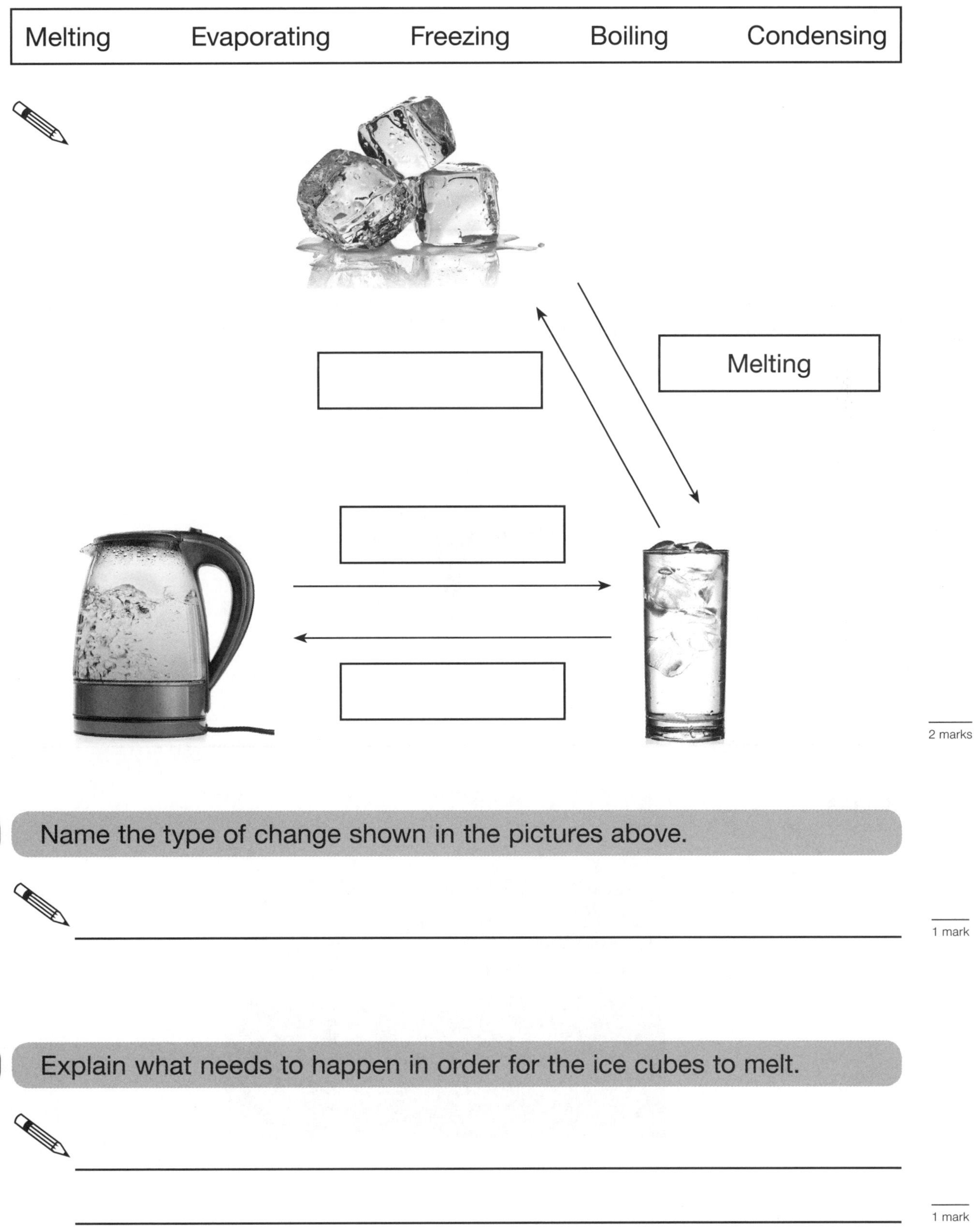

2 marks

b Name the type of change shown in the pictures above.

____

1 mark

c Explain what needs to happen in order for the ice cubes to melt.

____

____

1 mark

4 Materials

a What are the properties of these two materials?

Conductor ____________________

____

Insulator ____________________

____________________ 2 marks

b Give **ONE** important property of each of the following materials.

(i) Wood

____________________ 1 mark

(ii) Glass

____________________ 1 mark

(iii) Fabric

1 mark

c Explain the difference between transparent and opaque.

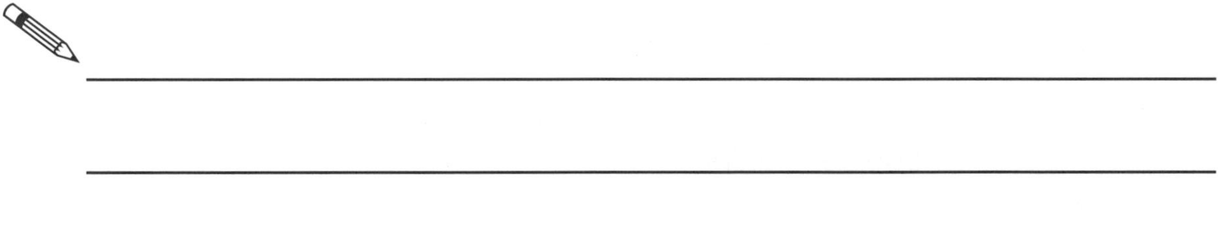

2 marks

d Copper is a good electrical conductor.

What does it mean if a material has a high value for electrical conductivity?

1 mark

5 Keeping warm

a Charlotte is trying to keep warm.

Name **TWO** items of clothing she can wear that could keep her insulated.

1. ______________________________

2. ______________________________

2 marks

b Charlotte wants to improve the length of time her house stays warm.

Which of the following would work? Tick all that apply.

- [ ] roof insulation
- [ ] double glazing
- [ ] wooden roof
- [ ] single glazing

1 mark

c A radiator is a thermal conductor.

What does this mean?

____

____

1 mark

Total ________/24 marks

Chemistry

- You have 25 minutes to complete this test.

TEST 2

1 Jelly

a Freya is investigating different ways to make jelly.

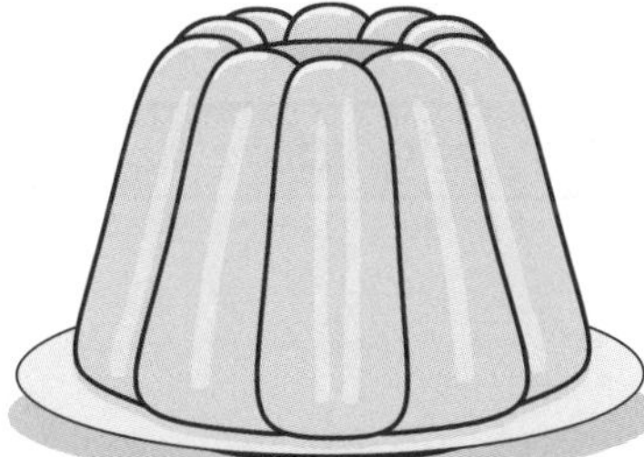

She is going to make two lots of jelly. First, she dissolves jelly cubes in water. She uses hot water for one lot of jelly cubes and cold water for the other.

Will the cubes dissolve more quickly in the hot water or the cold water?

____ 1 mark

b Freya notices that the jelly cubes don't look solid anymore.

What has happened to the jelly cubes in the water?

____ 1 mark

c When making her jelly, Freya wants to speed up the process.

What could Freya do to speed up the process that happens when she adds the jelly cubes to the water?
Explain your answer.

____

____

____ 2 marks

2 Solids, liquids and gases

a Jane is looking at materials. She needs to categorise the materials she is investigating.

Sort the following materials into the correct columns in the table below.

Steam	Water	Wood	Glass	Sand	Helium	Milk

Solids	Liquids	Gases

3 marks

b Give **TWO** properties that a material has when it is in each of the states below.

Solid

1. ______________________

2. ______________________

Liquid

1. ______________________

2. ______________________

Gas

1. ______________________

2. ______________________

3 marks

3 New bedroom

a Jed is planning a new bedroom. He has a variety of materials to use.

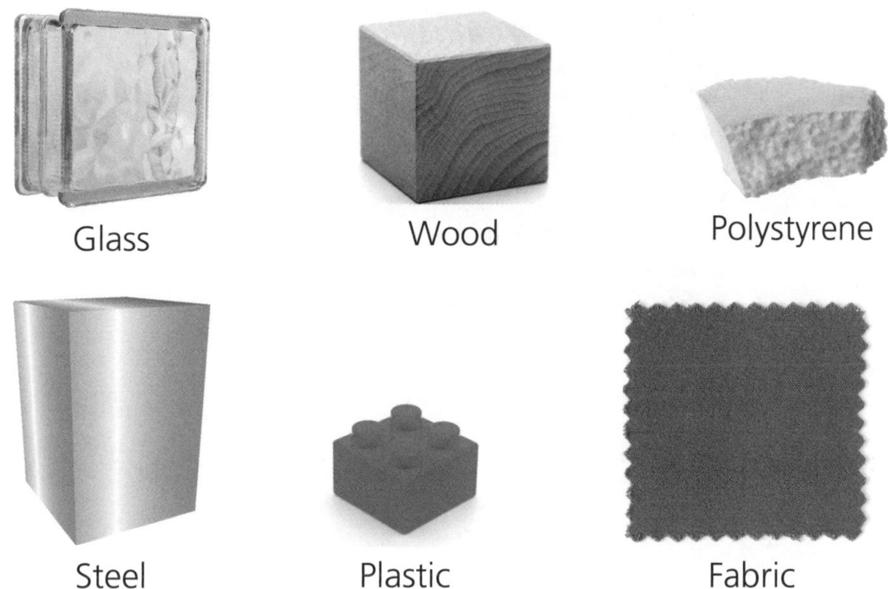

Jed needs a material to put into the windows in his room.

Which material would be best to use?

____ 1 mark

b Explain why this is the best material for windows.

____

____ 1 mark

c Which material is magnetic?

____ 1 mark

d Some fabrics are very absorbent.

What does this mean?

____

____

1 mark

e State what would need to be done to plastic to change its shape.

____

____

1 mark

f Plastic is a good electrical insulator.

What is the difference between an electrical insulator and an electrical conductor?

Electrical insulator ____

____

Electrical conductor ____

____

2 marks

4 Water cycle

a The water cycle is an extremely important natural event.

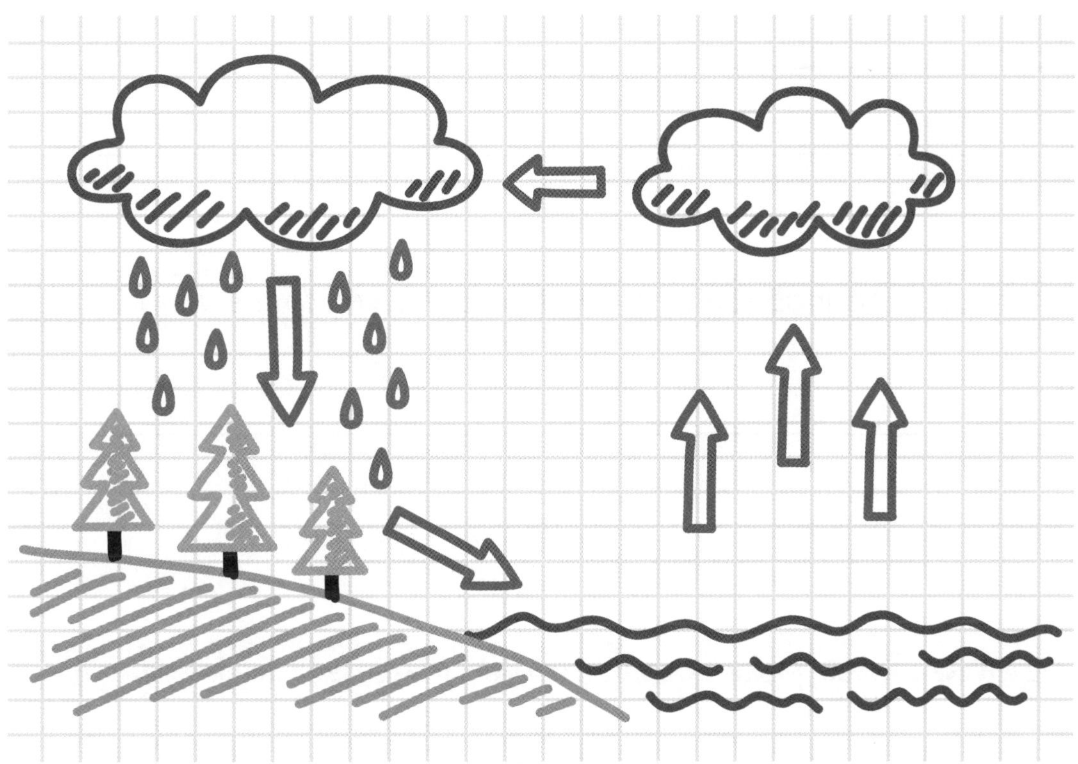

Write a number from 1 to 4 in each box to order these events in the water cycle.

One has been done for you.

1	Water evaporates into the air.
	Water falls as rain.
	Water vapour condenses into clouds.
	Water returns to the sea.

2 marks

b What makes the water evaporate?

____

____

____

2 marks

c When water vapour condenses, what is happening to the water?

____

____

1 mark

d Name **TWO** places water collects before it runs back to the sea.

1. _______________

2. _______________

2 marks

e In the water cycle, salty water from the sea eventually becomes rain water.

Why is rain water not salty?

____

____

1 mark

Total _______ / 25 marks

Chemistry

- You have 25 minutes to complete this test.

1 Rock and soil types

a Kelly has a number of rocks. She tries to scratch each rock with her fingernail. She can scratch off bits of the chalk but she cannot mark the other rocks.

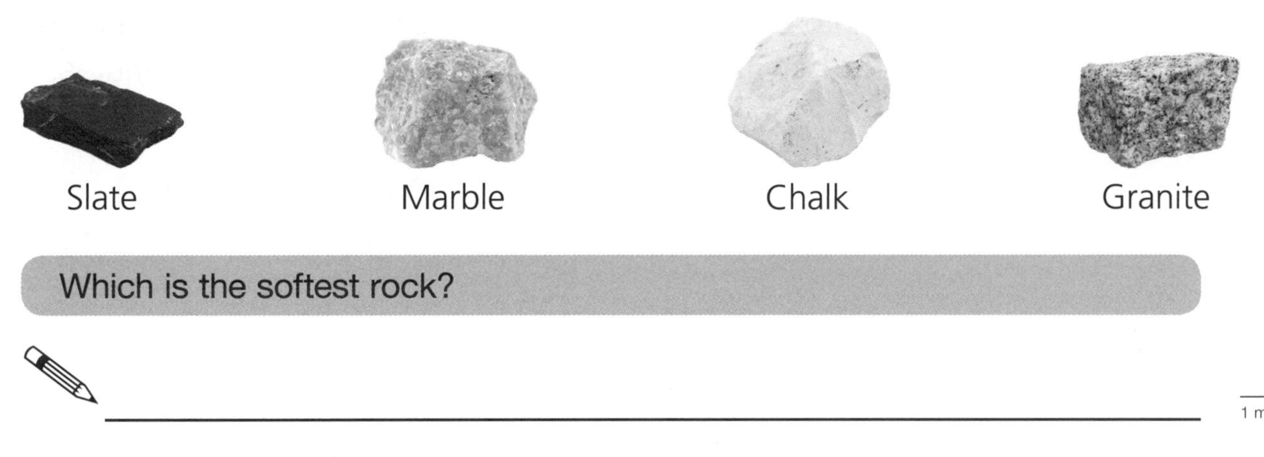

Which is the softest rock?

______________________________ 1 mark

b Kelly pours water onto each rock to see what will happen. Some rocks are permeable and others are impermeable.

What do these terms mean?

Permeable

____

____

____

Impermeable

____

____

______________________________ 2 marks

c Kelly is planting some plants to grow over the summer. She knows that certain plants will grow better in different soils.

Name **TWO** things that soil is made up of.

1. ______________________________

2. ______________________________ 2 marks

d Kelly puts different soils into funnels. She pours the same amount of water into each funnel and observes how much water comes through after 10 minutes.

Water

Soil

A B C

Which soil let the most water through?

______________________________ 1 mark

e Funnel A: When dry, clay soil has very few air gaps.
Funnel B: When dry, sandy soil has lots of air gaps.
Funnel C: Chalky soil allows water to drain through it quickly.

Which funnel has each of the different soils in it?

Clay ______________________________

Chalky ______________________________

Sandy ______________________________ 3 marks

2 Water temperature

a Jen is given four jars with water in them. She is told that some jars are warmer than others.

Fill in the table below with Jen's results.

Jar	Temperature (°C)
Boiling water	
Warm water	
Water with ice	
Hot water	

2 marks

b Jen puts some ice into each of the jars.

In which jar will the ice melt the fastest?

1 mark

c After the ice has melted Jen notices that the water level has increased from the original amount.

Why has this happened?

1 mark

d Jen takes the water from one jar and heats it in a pan on a hob.

Explain what will happen to the water.

____

____

____

2 marks

e What is the name of the process that makes steam change back to liquid water?

____

1 mark

f Jen puts another jar of water in the freezer.

At what temperature does water freeze?

____

1 mark

3 Changes in materials

a Sarah is making some toast.

Can Sarah turn her toast back into bread? Explain your answer.

____

____ 2 marks

b What kind of change is this?

____ 1 mark

c Tick which of the following are reversible changes.

Melting chocolate ☐

Baking bread ☐

Frying an egg ☐

1 mark

d The following diagrams are in no particular order.

(i) **Copy** the diagrams and add labelled arrows to show how ice, liquid water and steam are linked by reversible changes.

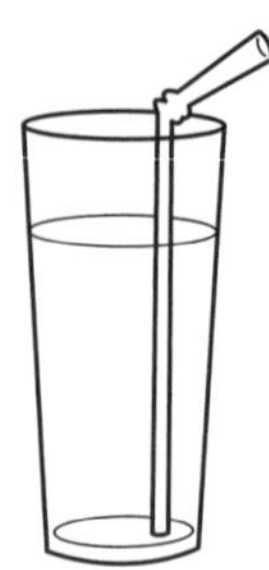

3 marks

(ii) Explain what your labelled diagram shows.

____

____

____

____

____

2 marks

Total ________/26 marks

Physics

TEST 1

- You have 25 minutes to complete this test.

1 Earth, Sun and Moon

a Joe is trying to find objects that show the Earth, Moon and Sun to scale for a project.

Write under each object which one of Earth, Moon and Sun would fit best.

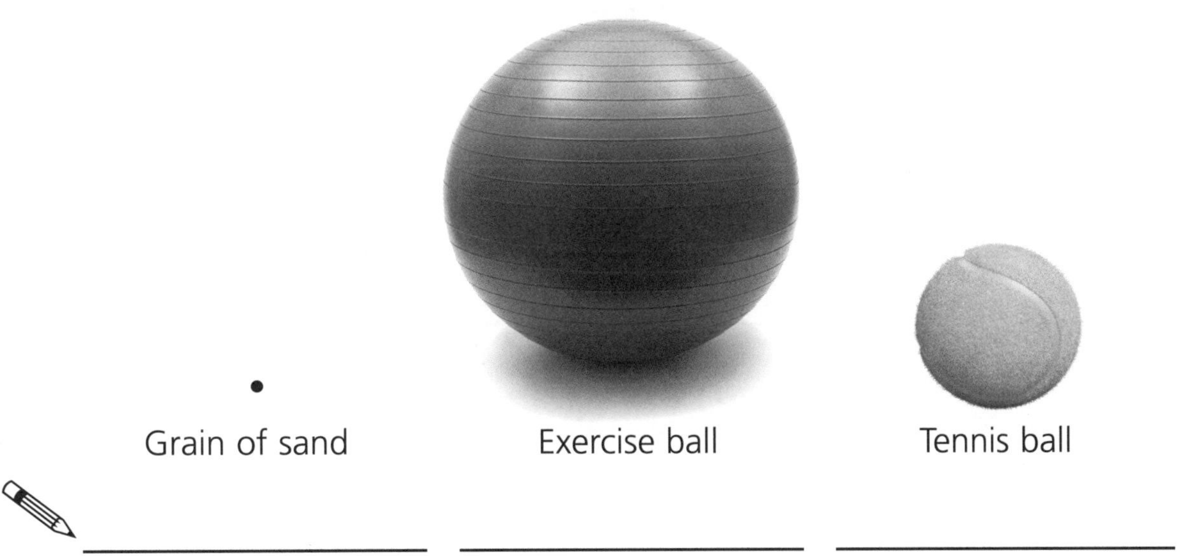

_______________ _______________ _______________

1 mark

b How often does the Moon orbit the Earth?

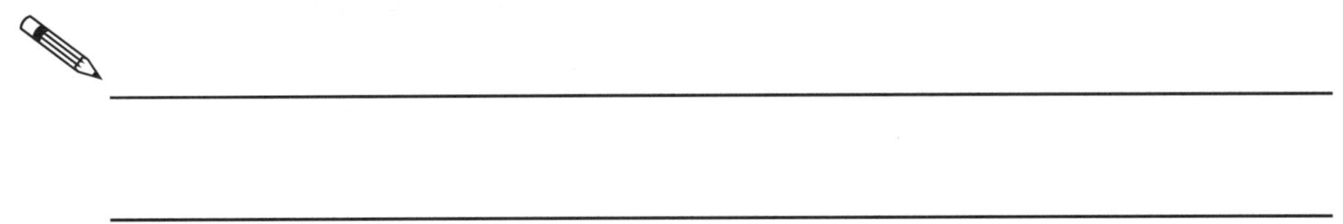

1 mark

2 Circuits

a Here is part of a series circuit.

(i) Finish drawing the circuit by including a closed switch and a bulb, using the correct symbols.

1 mark

(ii) Will the circuit work?

____

1 mark

b Will the lamp light in this circuit? Tick the correct answer box.

Yes ☐

No ☐

1 mark

c Will the motor work in this circuit? Tick the correct answer box.

M

Yes ☐

No ☐

1 mark

d Will the lamps light in this circuit? Tick the correct answer box.

Yes ☐

No ☐

1 mark

3 How we see

a Kamil is exploring how he can see different objects.

Using arrows, show how light travels so Kamil can see his scooter.

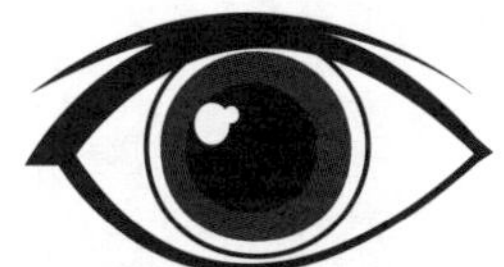

1 mark

b Give an example of a natural and an unnatural source of light.

Natural ______________________________

Unnatural ______________________________

2 marks

4 Friction

a Raaj has a toy car. He has been exploring how it travels on different surfaces and has noticed that it travels in different ways. He pushes the toy car and lets it go.

Put the following surfaces in order from the one on which the toy car travels the least distance to the one on which it travels the furthest.

_______________ _______________ _______________

Least Furthest

1 mark

b Explain why the toy car will eventually stop on the surfaces.

____

____

____

____

2 marks

5 Magnets

a Sort the following items into magnetic and not magnetic.

Gold	Steel	Plastic	Iron	Wood	Copper

Magnetic	Not magnetic

2 marks

b Explain why these magnets will not be attracted to each other.

S N N S

___

___

___

2 marks

c Complete the sentence below.

Sometimes magnets are not attracted to each other. In other words they ______________ each other.

1 mark

6 Making sounds

a Grace plucks the strings on the guitar and hears a sound.

What makes the sound?

____ 1 mark

b Grace can make sounds of different pitches.

Complete the following sentences.

Making a string shorter makes a sound with a ______________ pitch.

Making a string longer makes a sound with a ______________ pitch. 1 mark

c Grace plays a song for her friend.

Explain how Grace can make a soft sound and a loud sound on her guitar.

____

____

____

____ 2 marks

Total ________/22 marks

Physics

- You have 25 minutes to complete the test.

TEST 2

1 Gravity

a Ted drops a parachute from a 5 metre height. He knows that gravity and air resistance are acting on it.

Draw arrows on the diagram below to show the directions of these two forces.

2 marks

b Describe how the two forces act on the parachute.

Air resistance

____

____

Gravity

____

____

2 marks

c Ted has a second parachute. The second parachute is twice the size of the first.

Tick which parachute will fall more slowly.

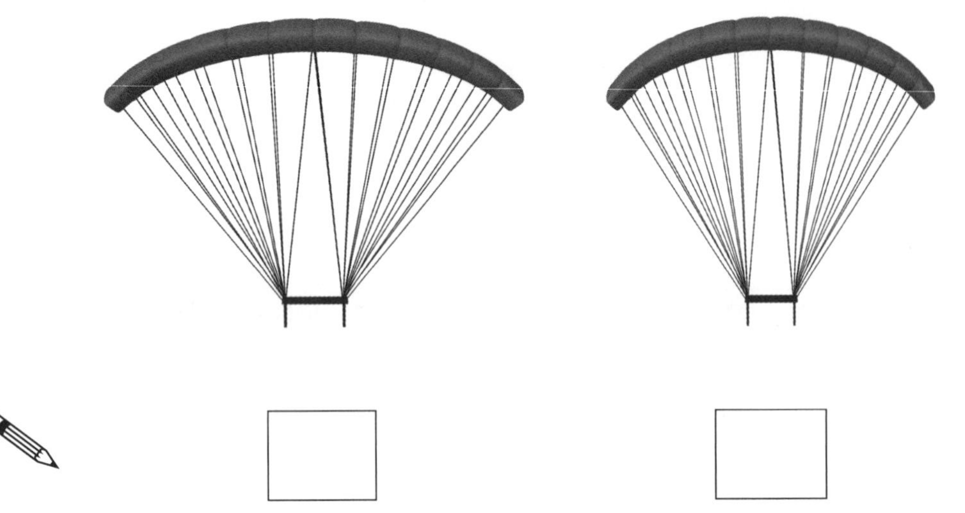

☐ ☐

1 mark

d Why does your chosen parachute fall more slowly?
Link your answer to air resistance.

____

____

____

______________________________ 2 marks

e Ted investigates gravity further. He has two pieces of paper. One is screwed up into a ball, the other is flat.

Which piece of paper will hit the floor first when they are dropped at the same time from the same height? Explain your answer.

______________________________ 2 marks

2 Forces

a Jay is collecting water from a well.

What system is in place to help Jay get the bucket up and down the well?

____

1 mark

b Name **ONE** other system that can be used to help people against forces.

____

1 mark

c Complete the following sentence.

Gears consist of a system of ____________ that allows a ____________

turning force to have a ____________ effect.

3 marks

3 Light and shadows

a Fred has a torch and is investigating how the position of the torch changes the shadow made against a block of wood.

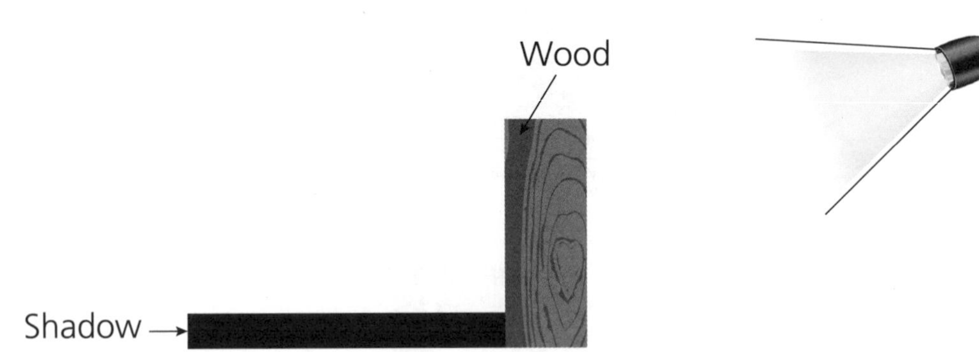

What will happen to the length of the shadow if the torch is lowered?

____

____ 1 mark

b Draw where the torch needs to be in order to create a shorter shadow than is shown in the diagram above.

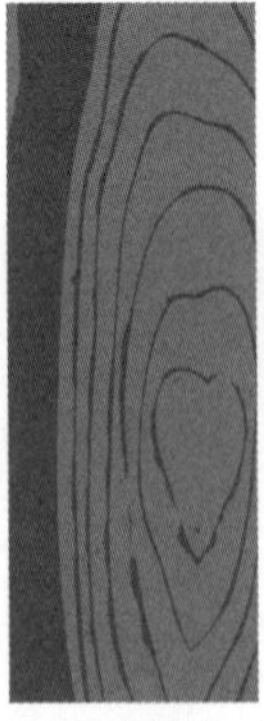

1 mark

4 Planets

a Put these planets in order from the closest to the furthest away from the Sun.

Jupiter

Neptune

Venus

Saturn

__________ __________ __________ __________

Closest | Furthest

2 marks

b Name the star at the centre of our solar system.

____

1 mark

5 How we see things

a Ruby is trying to look at a bowl of fruit placed behind a wall. She is investigating the ways in which she can see it.

Draw where Ruby could place a mirror in order to see the bowl with the wall in the way.

1 mark

b Explain how Ruby can see the bowl of fruit using a mirror.

____

____

____

1 mark

Total ______ /21 marks

Physics

- You have 25 minutes to complete this test.

TEST 3

1 Day and night

a Kim is making a note of where the Sun is at different times of the day. She comes up with three reasons for why the Sun moves throughout the day.

Which of these statements is true?

______________________________ 1 mark

b Explain how day and night occur in more detail.

____

____

______________________________ 2 marks

c Kim watches the Moon each night. She notices that the Moon looks different each night and on some nights she can't see it.

On the diagram draw the Moon and how it moves in relation to the Sun and Earth.

1 mark

d Kim comes up with four statements from her investigations on the Sun, Moon and Earth.

Mark if each statement is true (T) or false (F).

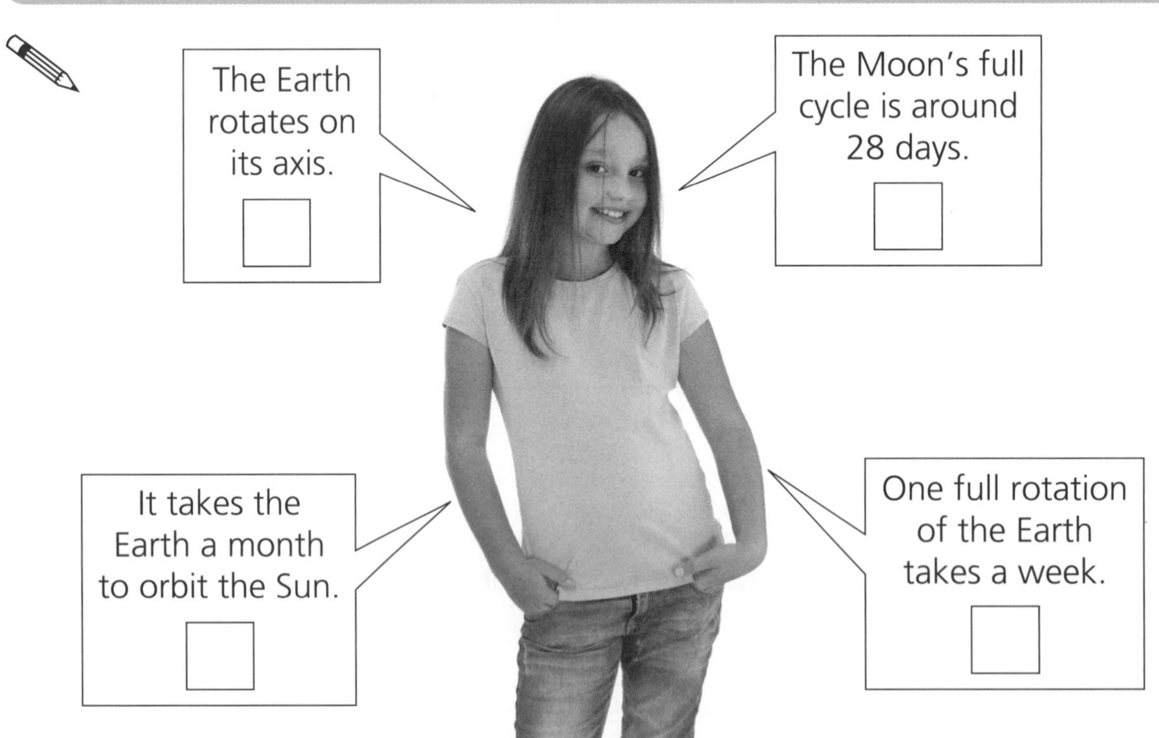

3 marks

2 Electrical objects

a Pryia is experimenting with some circuits. She is putting different objects in her circuits to see if the bulb will still light up.

(i) Put a tick in the correct box to indicate if the circuit will or will not work.

Iron bar

Will work ☐

Will not work ☐

1 mark

(ii) Put a tick in the correct box to indicate if the circuit will or will not work.

Plastic tube

Will work ☐

Will not work ☐

1 mark

b Pryia knows that there are electrical conductors and electrical insulators.

Explain what each of these are.

Conductor

____

____

Insulator

____

____

2 marks

c Pryia looks at some objects around her house. She organises them into two categories: runs on electricity and does not run on electricity.

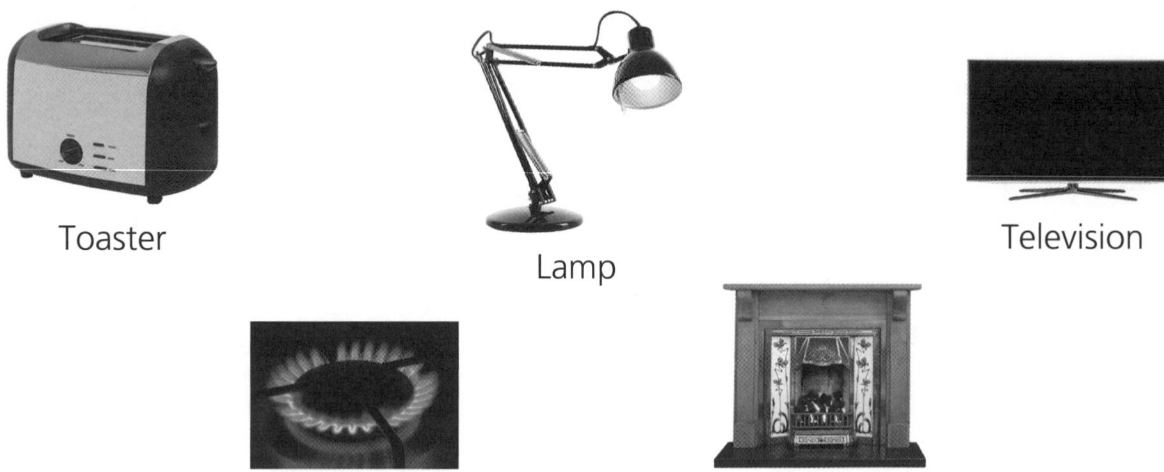

Sort the items into these two categories.

Runs on electricity	Does not run on electricity

2 marks

3 Spinners

a Katie is preparing her spinners for an investigation into forces. She is trying to think of something that she can change in her experiment. She decides to change the number of paper clips attached to the bottom of the spinner.

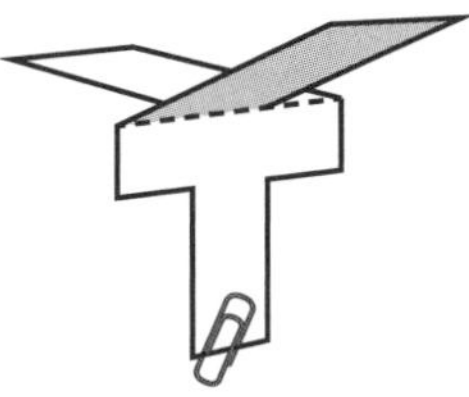

Write down TWO other changes she could have made.

1. ______________________________

2. ______________________________ 2 marks

b Katie performs the investigation and drops each spinner from the same height.

Which TWO main forces are acting on the spinner as it drops?

1. ______________________________

2. ______________________________ 2 marks

c Katie creates a table to show her results.

Number of paper clips	Time (s)
1	5.0
2	4.1
3	2.9
4	2.5

Katie dropped each spinner three times. She then took an average of the results.

Why does Katie need to drop each spinner three times and take an average?

____

______________________________ 1 mark

d Katie then creates a bar chart showing her results. Complete the bar chart.

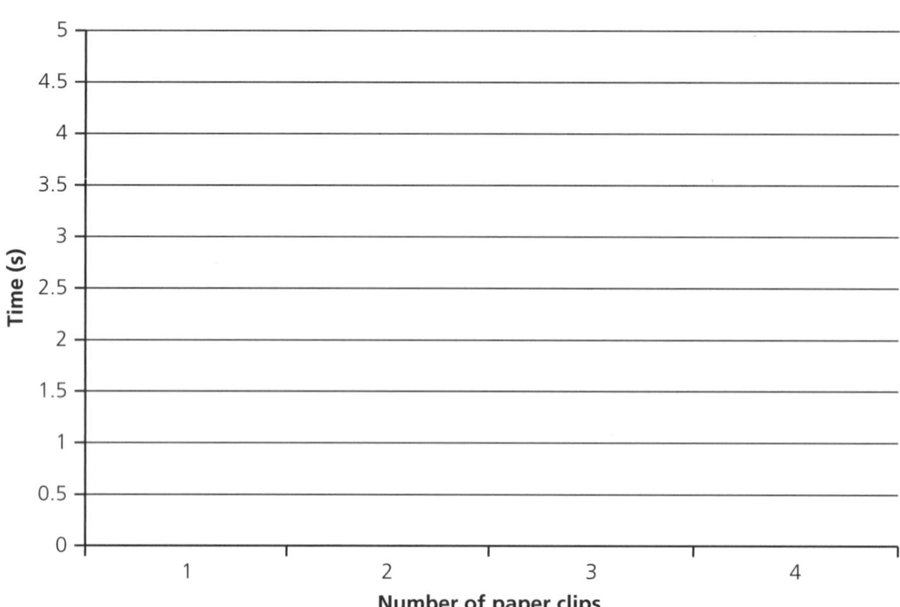

3 marks

e Katie uses her graph to interpret the results.

Which spinner fell the slowest?

____

1 mark

f Explain what would happen if there was a spinner with no paper clips.

____

____

____

1 mark

Total ________ /23 marks

Answers

Biology Test 1

1. a

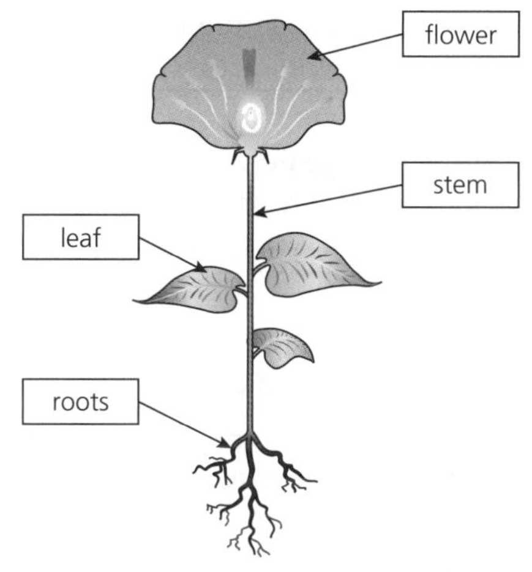

(1 mark)

b

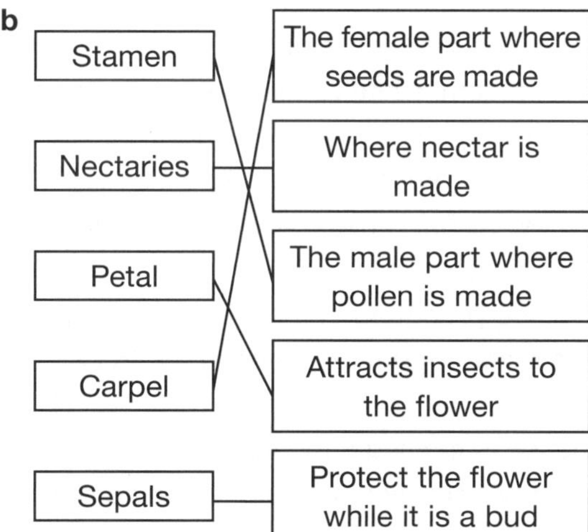

(3 marks: 0 marks for one correct answer, 1 mark for two or three correct answers, 2 marks for four correct answers, 3 marks for five correct answers)

2. a Canines – used for tearing and ripping food
Incisors – help to bite off bits of food
Molars – help to crush and grind food **(3 marks)**

b Any two from: brush teeth twice a day; avoid sugary food and drinks; visit the dentist regularly **(2 marks)**

3. a

(1 mark)

b One of the following for one animal:

Giraffe – can consume and store water for long periods of time; long necks to feed and see predators from a distance; tough tongues to pull leaves off without being hurt by thorns

Penguin – layer of fat and thick skin to keep warm; huddle together to keep warm; waterproof feathers help them to swim in cold water

Camel – long eyelashes to keep sand out of their eyes; nostrils that can open and close; thick eyebrows to shade the eyes from the sun; store water in their hump

Or any other suitable answers. **(1 mark)**

4. a Any three from: different sizes; different shapes; different colours; different fur; different ears **(3 marks)**

b Any two from: different eye colour; different hair colour; different feet size; different heights **(2 marks)**

5. a

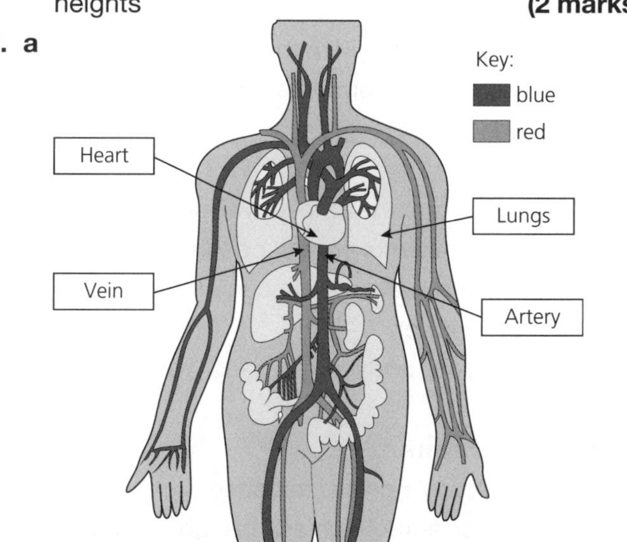

(3 marks: 1 mark for one or two correct answers, 2 marks for three correct answers, 3 marks for four correct answers)

b

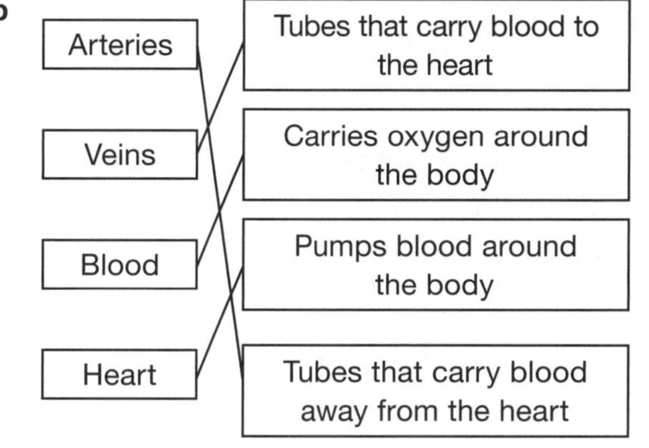

(2 marks: 0 marks for one correct answer, 1 mark for two or three correct answers, 2 marks for four correct answers)

6. **a** Any two from: to protect important organs in the body; to support the body; to help the body move **(2 marks)**

   b

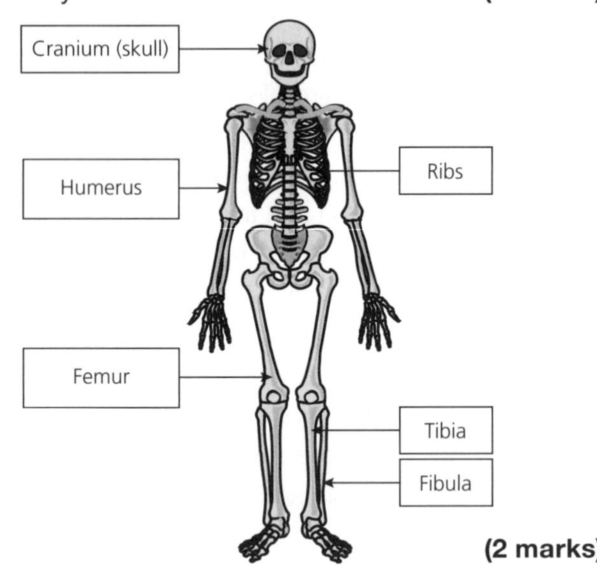

(2 marks)

Biology Test 2

1. **a** Any one from: to have a strong stem and keep the plant healthy; so that the plant can make its own food. **(1 mark)**
   b Water enters the plant through its roots. **(1 mark)**
   c It is transported up the stem to the leaves and flowers. **(1 mark)**
2. **a** Glen **(1 mark)**
   b Any one from: the smoker will struggle to breathe; the smoker may cough often; the smoker may get lung cancer; any other correct answer. **(1 mark)**
   c To help your body and organs work properly. **(1 mark)**
3. **a** The heart rate **(1 mark)**
   b The children's pulses increase **(1 mark)**
   c It is important to increase breathing so more oxygen is moving around the body. **(1 mark)**
   Your muscles are working harder so they need more oxygen. **(1 mark)**
4. **a** cabbage → slug → thrush **(1 mark)**
   b Cabbage – Producer
   Slug – Prey
   Thrush – Predator **(3 marks)**
5. **a**

3	Excess water is absorbed back into the body in the large intestine.
2	Digested food is absorbed into the bloodstream through the small intestine.
4	Any undigested food passes out when we go to the toilet.
1	Food is eaten, then digested in the mouth, stomach and small intestine.

(2 marks: 0 marks for one correct answer, 1 mark for two correct answers, 2 marks for four correct answers)

   b Oesophagus **(1 mark)**
6. **a** Any one from: white fur; layer of fat and thick layer of fur; wide large paws **(1 mark)**
   b Any one from: white fur helps the polar bear blend in with its surroundings; layers of fat and thick layers of fur keep the polar bear warm; wide paws help the polar bear walk on snow. **(1 mark)**
   c **1.** To see predators in the distance. **(1 mark)**
   2. To reach food that is high in the trees. **(1 mark)**
   d If the ice is melting it means the polar bears' habitat will slowly disappear, meaning polar bears may disappear one day. **(2 marks)**

Biology Test 3

1. **a** It is missing light **(1 mark)**
   b light, water, air **(3 marks)**
2. **a** Any one from: so they are able to stretch more easily; to avoid pulling a muscle when exercising. **(1 mark)**
   b When the biceps muscle contracts it will shorten and bring up the lower arm. **(2 marks)**
3. **a** Vertebrates have a backbone but invertebrates do not. **(1 mark)**
   b

Vertebrate	Invertebrate
mammals birds	spiders insects

(2 marks: 1 mark for each correct column)

c

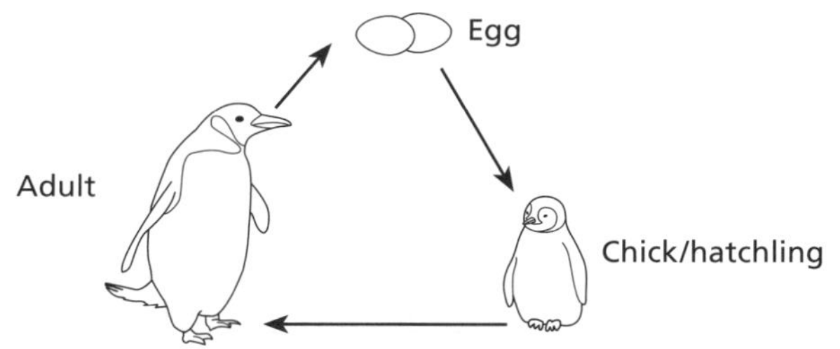

(1 mark)

4. a

3	Seeds are scattered by an animal or the wind.
1	Pollen is blown by the wind or carried by insects from one plant to another.
2	Pollen reaches the carpel of the new flower and travels to the ovary. It then fertilises ovules to make seed.

(2 marks: 1 mark for one correct answer, 2 marks for three correct answers)

b **(i)** Dispersal **(1 mark)**
(ii) Pollination **(1 mark)**
(iii) Fertilisation **(1 mark)**

c Petal – bright to attract insects
Stamen – the male part of the plant that makes the pollen
Carpel – the female part of the plant where the seeds are made **(3 marks)**

d This is when a new plant starts to grow from the seed. **(1 mark)**

e Roots **(1 mark)**

5. a Grows; eats; cries; urinates; drinks milk
(2 marks: 1 mark for one to three correct answers, 2 marks for four or five correct answers)

b

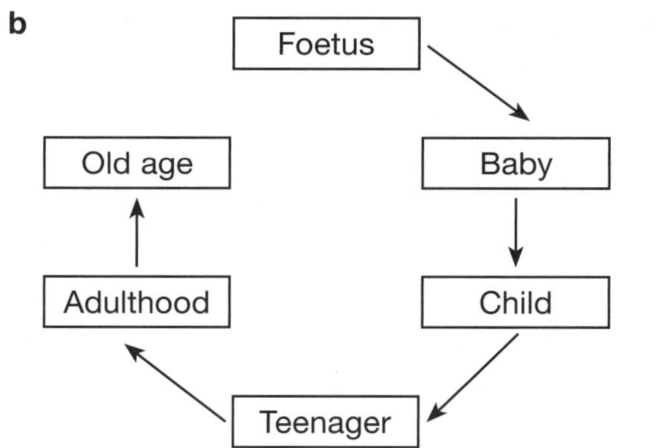

(2 marks: 1 mark for one or two correct answers, 2 marks for three or four correct answers)

Chemistry Test 1

1. a Ash, smoke, gases **(1 mark)**
b Irreversible **(1 mark)**

2. a Rock **(1 mark)**

b

3	Rock pushes the sediment down and water washes away the bones, leaving a space in the rock.
5	Fossils are uncovered millions of years later.
2	Sediment falls onto the skeleton of the animal.
1	The animal dies and drops to the river bed.
4	Water carries the rock into the area where the animal was, creating a fossil.

(3 marks: 1 mark for one or two correct answers, 2 marks for two to four correct answers, 3 marks for five correct answers)

c Any two from: how long the animal lived; how things are related; how the animal lived; the environment that they lived in. **(2 marks)**

3. a

(2 marks: 1 mark for each pair of correct answers)

b Reversible **(1 mark)**

c They need to have heat applied to them. **(1 mark)**

4. a Conductor – a material that allows electricity to go through it.
Insulator – a material that does not allow electricity to go through it; keeps something warm or cold **(2 marks)**

b (i) Wood – Strong
(ii) Glass – Transparent, brittle, electrical insulator
(iii) Fabric – Light, flexible
(3 marks: 1 mark for one correct answer for each material)

c Any two from: you can see through materials that are transparent; you cannot see through materials that are opaque; light can travel through transparent materials, but cannot travel through opaque materials.
(2 marks: 1 mark for each correct definition)

d Electricity travels through it easily **(1 mark)**

5. a Any two from: jumper; coat; socks; scarf; gloves; hat **(2 marks)**
b roof insulation, double glazing **(1 mark)**
c It allows heat energy out into the room. **(1 mark)**

Chemistry Test 2

1. a Hot water **(1 mark)**
b They have dissolved. **(1 mark)**
c She could use hotter water and stir the jelly. **(1 mark)**
This will make the jelly cubes dissolve more quickly. **(1 mark)**

2. a Solids – wood, glass, sand
Liquids – water, milk
Gases – steam, helium
(3 marks: 1 mark for two or three correct answers, 2 marks for four or five correct answers, 3 marks for six or seven correct answers)

b Solid Stays in one place and can be held; keeps its shape; can be cut or shaped
Liquid Can flow and be poured; changes shape to fit its container
Gas Does not keep its shape; fills the space it is in; can be squashed
(3 marks: 1 mark for two correct statements for each state)

3. a Glass **(1 mark)**
b Glass is the best material for windows because it is transparent. **(1 mark)**
c Steel **(1 mark)**
d It means that fabric holds water when it gets wet. **(1 mark)**
e Plastic needs to be heated, then it can be moulded. **(1 mark)**
f Electrical insulator means that electricity cannot pass through a material. Electrical conductor means that electricity can pass through a material. **(2 marks)**

4. a

1	Water evaporates into the air.
3	Water falls as rain.
2	Water vapour condenses into clouds.
4	Water returns to the sea.

(2 marks: 1 mark for one or two correct answers, 2 marks for three or four correct answers)

b The sun heats the water up. **(1 mark)**
This creates water vapour. **(1 mark)**
c It is changing from vapour to liquid. **(1 mark)**
d 1. Lakes
2. Rivers **(2 marks)**
e Any one from: water evaporates but salt does not; only water evaporates. **(1 mark)**

Chemistry Test 3

1. a Chalk **(1 mark)**
b Permeable – A permeable rock allows water to soak through it
Impermeable – An impermeable rock does not allow water to soak through it. **(2 marks)**
c Any two from: tiny pieces of rock; tiny pieces of dead plants; tiny pieces of dead animal; tiny pieces of air; tiny drops of water **(2 marks)**
d B **(1 mark)**
e Clay – C
Chalky – B
Sandy – A
(3 marks: 1 mark for each correct answer)

2. a Boiling water – 100°C
Warm water – 20°C
Water with ice – 5°C
Hot water – 50°C
(2 marks: 1 mark for one or two correct answers, 2 marks for three or four correct answers)
b The jar containing boiling water. **(1 mark)**
c The water from the melted ice has added to the water that is already in the jar. **(1 mark)**
d It will boil and evaporate **(2 marks)**
e Condensation **(1 mark)**
f 100°C **(1 mark)**

3. a No because it has been cooked **(2 marks)**
b Irreversible change **(1 mark)**
c Melting chocolate **(1 mark)**

d (i)

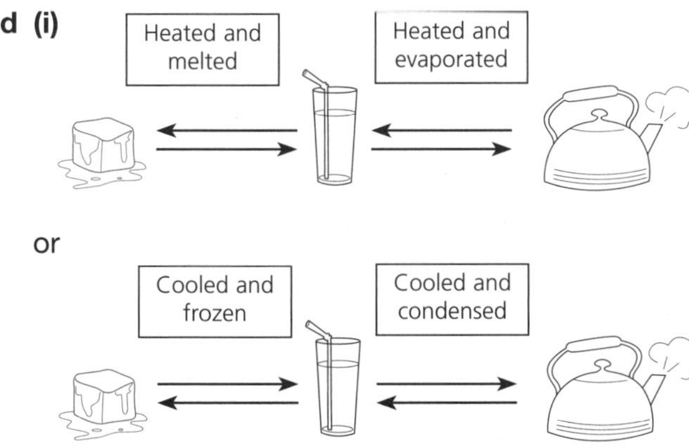

or

(3 marks: 1 mark for correct arrows, 1 mark for correct vocabulary, 1 mark for correct order)

(ii) If the ice is heated it will melt into a liquid. If the water is heated the liquid will turn into a gas and evaporate. If the steam is cooled it will condense and turn back into water. If the water is cooled again it will freeze and turn into ice. **(2 marks)**

Physics Test 1

1. **a** Grain of sand – Moon
   Exercise ball – Sun
   Tennis ball – Earth **(1 mark)**

   b Any one from: every 28 days; every month; every four weeks **(1 mark)**

2. **a (i)**

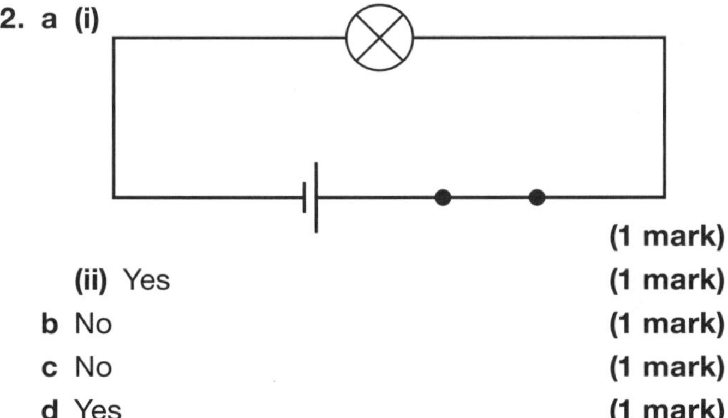

(1 mark)

   (ii) Yes **(1 mark)**

   b No **(1 mark)**

   c No **(1 mark)**

   d Yes **(1 mark)**

3. **a**

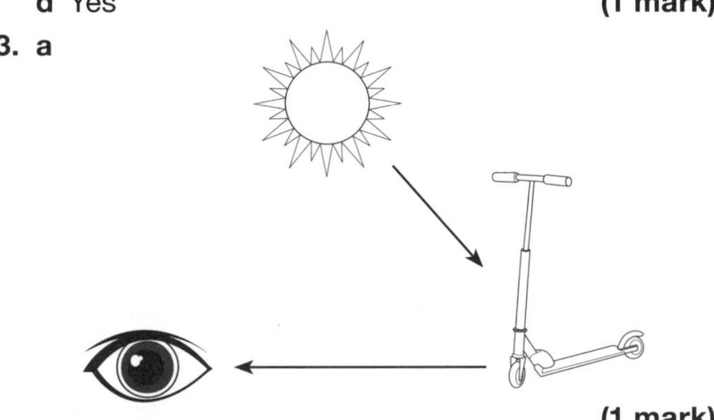

(1 mark)

   b Natural – one from Sun, candle, oil burner, fire
   Unnatural – one from electric light, television **(2 marks)**

4. **a** Long grass, carpet, tiles **(1 mark)**

   b Eventually the car will stop on all the surfaces because of friction. Friction is caused when two surfaces slide across each other. **(2 marks)**

5. **a**

Magnetic	Not magnetic
steel iron	gold copper plastic wood

(2 marks: 1 mark for each correct column)

   b The magnets will not be attracted to each other because the two north ends are facing each other. The north end needs to face the south end to attract. **(2 marks)**

   c Repel **(1 mark)**

6. **a** The sound is made through vibrations **(1 mark)**

   b higher
   lower **(1 mark)**

   c A soft sound is created by plucking gently on the guitar string. **(1 mark)**
   A louder sound is made by plucking the guitar string harder. **(1 mark)**

Physics Test 2

1. **a**

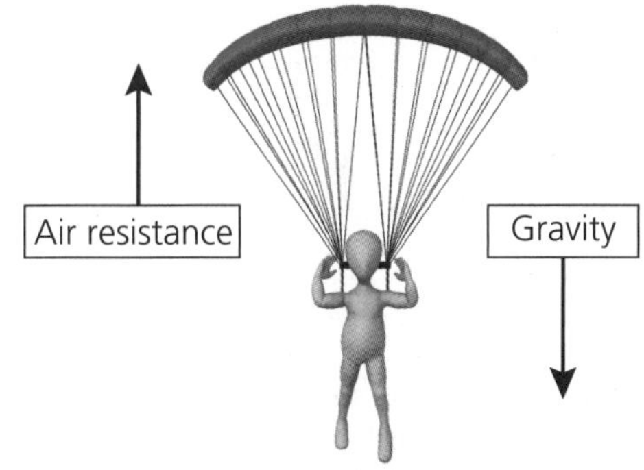

(2 marks: 1 mark for each correct arrow)

   b Air resistance – makes the parachute slow down
   Gravity – pulls the parachute to the ground/pulls the parachute towards the Earth
   (2 marks: 1 mark for each correct answer)

   c

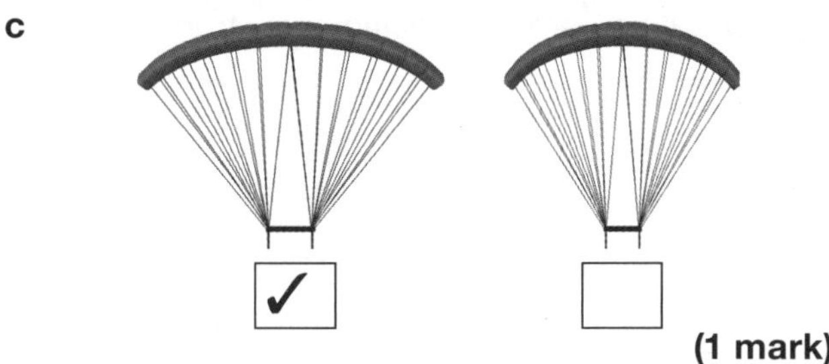

(1 mark)

   d The parachute on the left has a greater surface area, so the air resistance is greater for this parachute. **(2 marks)**

e The screwed up piece of paper will hit the floor first. It has less surface area than the flat piece of paper. **(2 marks)**

2. a Pulley system **(1 mark)**

b Lever system **(1 mark)**

c Gears consist of a system of *cogs* that allows a *small* turning force to have a *great* effect. **(3 marks)**

3. a It will increase in length. **(1 mark)**

b

(1 mark)

4. a Venus, Jupiter, Saturn, Neptune

(2 marks: 1 mark for one to three correct answers, 2 marks for four correct answers)

b The Sun **(1 mark)**

5. a

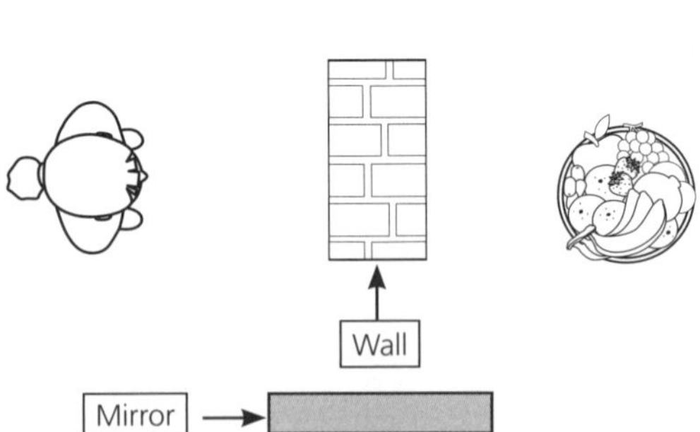

Mirror can be shown in either position. **(1 mark)**

b A mirror reflects light, meaning the image of the bowl of fruit will reflect in the mirror and Ruby can see it. **(1 mark)**

Physics Test 3

1. a 3. The Earth must orbit the Sun! **(1 mark)**

b The Earth rotates on its axis. When the place you are at is facing the Sun, it is light. When the place you are at is facing away from the Sun, it is dark. **(2 marks)**

c

(1 mark)

d The Earth rotates on its axis. T

It takes the Earth a month to orbit the Sun. F

The Moon's full cycle is around 28 days. T

One full rotation of the Earth takes a week. F

(3 marks: 1 mark for one or two correct answers, 2 marks for three correct answers, 3 marks for four correct answers)

2. a (i) Will work **(1 mark)**

(ii) Will not work **(1 mark)**

b Conductor – a material that will let an electrical current through.

Insulator – a material that will not let an electrical current through.

(2 marks: 1 mark for each correct answer)

c

Runs on electricity	Does not run on electricity
toaster lamp television	coal fire gas hob

(2 marks: 1 mark for each correct column)

3. a Any two from: length of the spinner's wings; width of the spinner's wings; material used to make the spinner; any other sensible change

(2 marks: 1 mark for each change)

b Air resistance

Gravity

(2 marks: 1 mark for each correct answer)

c Dropping the spinners three times and taking an average improves the accuracy of the test. **(1 mark)**

d

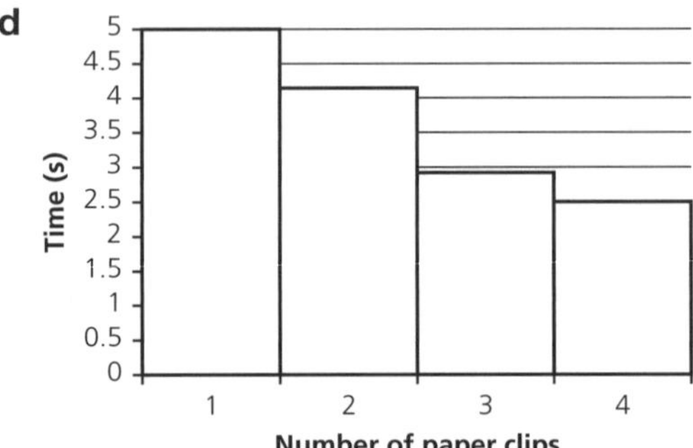

(3 marks: 1 mark for one or two correct answers, 2 marks for three correct answers, 3 marks for four correct answers)

e The one with one paper clip. **(1 mark)**

f As there are no paper clips on the spinner, it will fall more slowly than the spinners that have paper clips on because there is less weight. **(1 mark)**

Progress Report

Fill in your total marks for each completed test.
Colour the stars to show how you feel after completing each test.

☆ = needs practice ☆☆ = nearly there ☆☆☆ = got it!

Biology

Test	Marks	How do you feel?
Test 1	/ 25	☆ ☆ ☆
Test 2	/ 23	☆ ☆ ☆
Test 3	/ 25	☆ ☆ ☆

Chemistry

Test	Marks	How do you feel?
Test 1	/ 24	☆ ☆ ☆
Test 2	/ 25	☆ ☆ ☆
Test 3	/ 26	☆ ☆ ☆

Physics

Test	Marks	How do you feel?
Test 1	/ 22	☆ ☆ ☆
Test 2	/ 21	☆ ☆ ☆
Test 3	/ 23	☆ ☆ ☆